THE BIRTH OF OUR PLANET

THE EARTH, ITS WONDERS, ITS SECRETS

THE BIRTH OF OUR PLANET

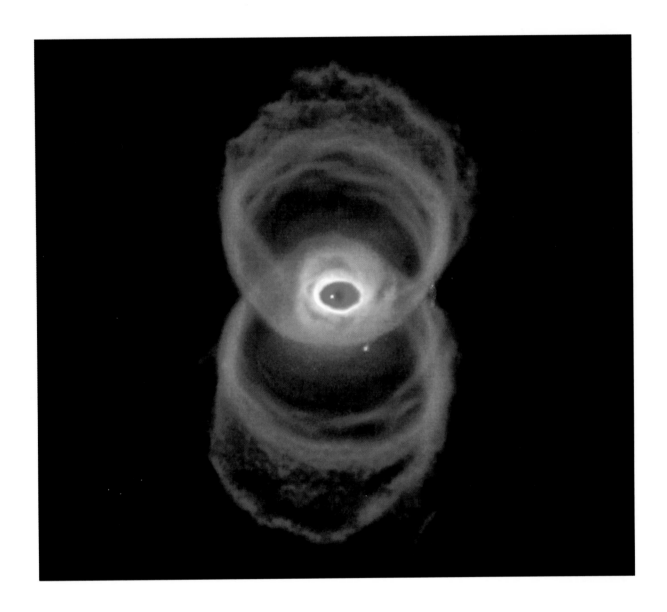

Reader's Digest

PUBLISHED BY

THE READER'S DIGEST ASSOCIATION LIMITED

LONDON NEW YORK MONTREAL SYDNEY CAPE TOWN

THE BIRTH OF OUR PLANET
Edited and designed by Toucan Books Limited
with Bradbury and Williams
Written by John Man
Edited by Helen Douglas-Cooper
Picture Research by Marian Pullen

FOR READER'S DIGEST, UK
Series Editor: Christine Noble
Editorial Assisitant: Chloe Garrow
Editorial Director: Cortina Butler
Art Director: Nick Clark

FOR READER'S DIGEST, U.S.
Senior Editor: Fred DuBose
Design Director: Irene Ledwith
Group Editorial Director, Nature: Wayne Kalyn
Vice President, Editor-in-Chief: Christopher Cavanaugh
Art Director: Joan Mazzeo

First English Edition Copyright © 1997
The Reader's Digest Association Limited,
11 Westferry Circus, Canary Wharf, London E14 4HE

Copyright © 1997 Reader's Digest Association, Inc.
Copyright © 1997 Reader's Digest (Canada), Inc.
Copyright © 1997 Reader's Digest (Australia) Pty Limited
Copyright © 1997 Reader's Digest Association Far East Limited
Philippines copyright © 1997 Reader's Digest Association Far East
Limited

Reprinted with amendments 1999

Separations: David Bruce Graphics Limited, London

Printed in the United States of America, 1999

Library of Congress Cataloging in Publication Data has been
applied for.

ISBN 0-7621-0139-3

FRONT COVER *An optical image of the Orion nebula. A satellite image
of the whole Earth (inset).*

PAGE 3 *The hourglass-shaped nebula known as MyCn18, photographed
from the Hubble Space Telescope.*

Contents

THE UNIVERSE'S RECIPE FOR LIFE

In a Universe where everything, from the smallest dust particles to the largest stars, has been formed from the same ingredients and is governed by the same laws, at least one planet has evolved with the conditions necessary for life.

In the early 1970s, a popular poster showed the Earth, marbled with the whites and blues of cloud and ocean, rising above the dead rim of the Moon. The photograph, taken by Neil Armstrong during the first manned landing on the Moon in 1969, became an international icon, for it was the first time that anyone had seen the Earth in this way and from so far away. What they saw in that image was a living world contrasted with a dead one, set against a dark and hostile Universe. Suddenly, the world seemed a rare and precious place.

The sight was both a shock and a revelation, because it provided a new way of seeing what we already knew and took for granted. The shock was also one of recognition. So that's what home looks like!

What, though, if we strip away the intensely human reactions to the image. What if a person had no feeling that this planet was home? What if that marbled globe could be seen through the eyes of explorers from some alien civilisation? Would there be any sense among them that this third planet from a rather ordinary star (our Sun) is extraordinary? The answer is yes, without a doubt.

Of course, if aliens arrived right now, their deductions would be made easy by the flow of broadcast information and by the sight of urban areas lit up at night. But, since this is a thought experiment, and setting aside the practicalities of interstellar space travel, suppose aliens had arrived before the development of industrial society, at any time before 1800. What would they deduce?

AN ALIEN ASSESSMENT

Even from as far away as the Moon, interstellar visitors would know that the Earth is a significant and interesting place. The planet before them would be in just the right position for varied and complex chemical reactions – not too far from the Sun to be frozen, not too close to be fried. It would be large enough for gravity to hold its volatile atmospheric gases in place. It revolves as it orbits the Sun, which would create a complex range of temperatures, with effects that soon become obvious. As the aliens approach, warily, they would notice that some of the colours change over time. This is a world with seasons. Other planets in this solar system have seasons, but nothing like this, for here is a world seething with change, rich with the potential for life.

The aliens would record the ice-covered poles and the oceans, and deduce that the temperature is around the crucial 100°C (212°F) point that allows hydrogen and oxygen to combine as water and circulate freely. They would know how easily this balance can be upset, and what the consequences would be: a world either seared by heat or frozen to death. They might

DISTANT PERSPECTIVE
The living Earth comes into view as it rises above the horizon of the barren Moon.

comment on the greens of tropical vegetation and the arid browns of deserts, and realise that there is a good range of soils and climates. They would see great rivers flowing from mountain ranges, and swirling cloud formations, and know that water must be seized by winds, form clouds and fall again as rain.

They would realise, too, that the planet's dynamism is not a recent development. It is a restless place, through and through. The edges of the continents match each other, like pieces of a scattered jigsaw. Initially, perhaps, the alien observers might guess that the land is actually floating, until they note massive features, such as the Great Rift Valley, along which the land is being unzipped by some titanic, slow-acting subterranean force. They would identify other types of cloud, effluvia from erupting volcanoes, and see them as evidence of hidden forces that might explain how the landmasses are being torn apart.

Wind, ocean, land and ice; forests, grasslands, uplands, deltas, tundra and deserts: all interact. It is as if the Earth is a giant test tube constantly being shaken up, yet constantly preserving stability – just the sort of environment that produces ever more complex chemical and biochemical combinations. As the alien explorers gather information, it would begin to link up, suggesting mechanisms that explain how this seething, evolving world worked. Moreover, the large quantity of greenery would point to the most fascinating complexity of all: life. And where there is life in one form, there could well be life in more advanced forms. This world would surely fascinate any creature intelligent and curious enough to undertake space travel.

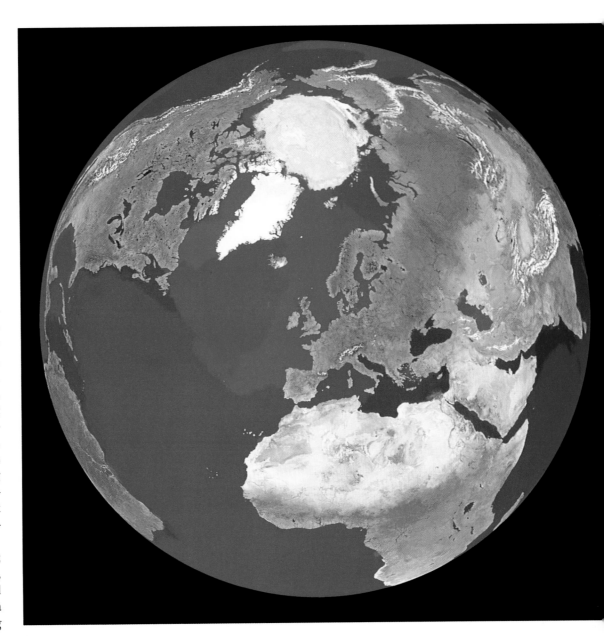

In addition, the chances are that the alien space crew would have a double shock, because Earth would remind them of their own home. It would not matter where they came from. The Universe is a uniform place, made of the same material everywhere and governed by the same laws. Sun-like stars are common enough. If any of them have planets – and there are good reasons for thinking that some do – and if they are life-bearing planets, almost certainly they will share characteristics with the Earth. Whatever the aliens' form, whatever shape the random forces of evolution have

VITAL PLANET *The Earth's landmasses and oceans are indications that active forces are at work on the planet, suggesting the possibility of life.*

given them, their planet is likely to be comparable in size to Earth, about the same distance from its sun, with an atmosphere, land, free-flowing water, and a rich and varied range of creatures.

This conclusion follows from the uniformity of the Universe. All stars, planets and life forms share a common heritage rooted

HEAVENLY STUDIES *From the earliest times, humans have studied the stars. This celestial planisphere dated 1708 shows the signs of the zodiac.*

in the beginning of the Universe, an event that is generally accepted to have occurred about 15 billion years ago. According to this theory – itself one of the most astonishing intellectual developments in human history – the matter and radiant energy that form most of the Universe in its current state came into existence at that moment. The simple raw material then underwent a long evolution, processing itself into ever greater complexity, until worlds as varied and beautiful as ours became a possibility.

Like every aspect of this restless Universe, all the Earth's marbled beauty, all its millions of life forms, all the random events of human history, all our creativity and yearning, spring ultimately from the fires of creation. In the words of one of the pioneers of modern astronomy, Sir Bernard Lovell, who devised the great grey bowl of the Jodrell Bank Radio Telescope that looms up over the Cheshire countryside: 'Is

it not extraordinary that the possibility of talking here this afternoon depends on events that were narrowly determined over 10 billion years ago?'

SIGNS OF LIFE

Having realised the commonality of origin, aliens approaching the Earth at any time in the last millennium would have seen signs of intelligent life – a great city, perhaps, or the Great Wall of China, or pyramids in the Egyptian desert. If a visit coincided with the late 20th century, they would know that intelligent life had begun to move beyond its planetary confines. Even a single piece of space junk (and there are some 6000 bits of discarded rocketry in Earth-orbit) would speak volumes. If they were lucky enough to intercept the Pioneer 10 or 11 spacecraft, they would receive a sudden insight into the nature of human intelligence. In any event, they would deduce that there existed on the planet below them a society with technological

know-how, a sophisticated social structure, restless curiosity and a great range of scientific acumen, including accurate astronomical knowledge.

The progress of human knowledge would be clear by implication. Earth's inhabitants would be skilful enough to gain an understanding of gravity, and of the nature of the Solar System. They would know how to turn raw materials into complex solids and liquids. They would understand how the Sun produced its energy, the nature of the stars, the structure of the Universe. Such knowledge demands sophisticated Earth-based instruments. Earth's inhabitants would know that space ships and space suits would be enough to provide an atmosphere in which they could breathe. Now that the first great hurdle had been leaped – placing a space vehicle in Earth-orbit – new chapters of manned and unmanned space travel would soon open.

The realisation of all this could well be enough to give any wise alien explorer pause for thought, for they would have no knowledge of the motives that have driven humans to explore the nature of their

MOON LANDING *Astronauts on the Apollo 15 mission prepare to explore the Moon in the Lunar Roving Vehicle.*

MARTIAN MONTAGE
Our planetary neighbour, Mars, has been photographed by the unmanned Viking I and II probes.

(1923), was a masterpiece of visionary theory and down-to-earth practicality.

RACING INTO SPACE

It was no coincidence that these advances, all inspired by the purest of scientific motives, were soon taken up for other, baser reasons. Oberth's work was continued by his assistant, Wernher von Braun, who worked on the development of a long-range rocket during the 1930s, culminating in the launch of the A-4 – renamed the V2. The V2 was a ballistic missile that could reach a height of 60 miles (96 km) – over half way to a near-Earth orbit – and it became the inspiration for the intercontinental ballistic missiles developed by both the USA and USSR in the 1950s as a means of delivering nuclear bombs. In 1957, the Soviet Union launched the first artificial satellite, Sputnik 1, into Earth's orbit, and in 1961 a Soviet pilot, Lt Yuri Gagarin, was the first person into space, orbiting the Earth in Vostock 1. In response, the Americans resolved to be the first to put a man on the Moon with their Apollo missions in the 1960s. All these missions spun off untold scientific and technological benefits, but it is unlikely that the immense amounts of money would have been found, or that the focus would have been so firmly on manned flight, without the spur of superpower rivalry.

As the relationship between

POPULAR VISITOR *In 1997 the comet Hale-Bopp, a visitor from the farthest regions of the Galaxy, was visible from Earth for several weeks.*

Universe. Human motives have been an odd mixture, reflecting the way science and history interweave.

At first, there was pure curiosity, epitomised by three men, a Russian, an American and a German. A century ago, when humans had still to conquer the air, a partially deaf Russian schoolteacher, Konstantin Tsiolkovsky, was already dreaming of the stars. In 1883, his first article on space flight, 'Free Space', accurately described weightlessness in space. In his *Dreams of Earth and Sky* (1895), he wrote of an artificial Earth satellite. His pioneering calculations proving that space flight by rocket was possible were published in 1903, the year the Wright brothers first took to the air. He even proposed a fuel – liquid hydrogen and liquid oxygen – which would later propel Saturn V to the Moon. In the USA, Robert Goddard, a Massachusetts physics teacher, independently proposed space travel in a pamphlet, *A Method of Reaching Extreme Altitude* (an understatement if ever there was one: by 'extreme altitude' Goddard meant the Moon). In 1926, with a Smithsonian Institution grant, Goddard launched the world's first liquid-fuelled rocket; it travelled 180 ft (55 m) and landed in his aunt's cabbage patch. No one saw much use for his invention. One sneering headline read: 'Moon Rocket Misses Target by 238 799.5 miles [384 300 km].' In Germany, Hermann Oberth, another teacher, came to the same conclusions as Tsiolkovsky and Goddard. His book, *The Rocket into Interplanetary Space*

the Soviet Union and the United States improved, the political inspiration for spectacular manned space missions vanished. A new era opened in which scientists found ingenious ways of cashing in on space technology with cheaper, smaller-scale, automated missions. The basis for such ventures was laid in the 1970s with probes to the outer planets, which took years – decades even – to produce results. Voyagers 1 and 2, launched in 1977, took two years to reach Jupiter; Voyager 2 continued its journey, flying past Neptune ten years later. Pioneer 10, launched in 1972, became the first manmade object to leave the Solar System, in 1987, and broadcast minute bleeps of information until it was finally abandoned to interstellar space in 1997. Probes have mapped Mercury, Venus, Mars and several satellites. Future missions will re-examine Mars for signs of life, while others will look for life on Jupiter's moons and examine the nature of Saturn's rings.

Such work can only be done close up. But other types of knowledge can increasingly be gleaned by ever more sophisticated instruments, either on Earth or in orbit.
continued on page 12

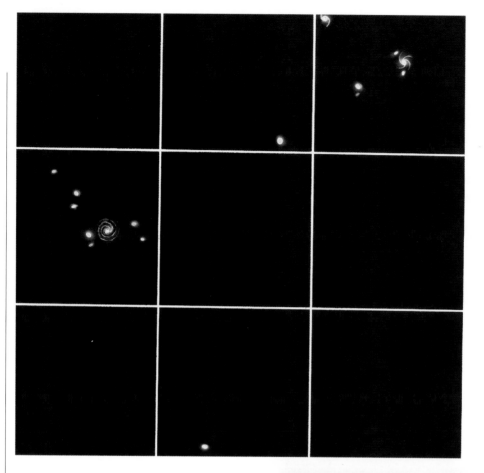

OPPOSITE: THE SOLAR SYSTEM
*Nine planets and their moons,
together with comets,
asteroids and general debris,
orbit the Sun, held in place by
the Sun's gravitational field.
Moving out from the Sun, the
planets are Mercury, Venus,
Earth, Mars, Jupiter, Saturn,
Uranus, Neptune and Pluto.
Each section of the grid
represents approximately
2 billion miles (3 billion km).*

THE LOCAL GROUP OF GALAXIES
*Our Galaxy, the Milky Way
(located in the centre of the
centre left section of the grid),
belongs to a small
cluster of galaxies known as
the Local Group. Each section
of the grid represents 1 million
light years (approximately
6 billion billion miles/10 billion
billion km). The observable
Universe has about 10^{12} or
1000 billion times the volume
of this cluster of galaxies.*

THE MILKY WAY *The Milky
Way is a spiral galaxy made
up of hundreds of billions of
stars. Our Solar System is
located in an outer arm (in the
top centre section of the grid).
Each section of the grid
represents 40 000 light years
(approximately 250 million
billion miles/400 million
billion km).*

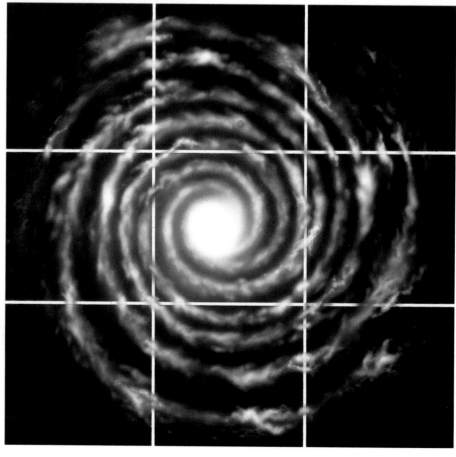

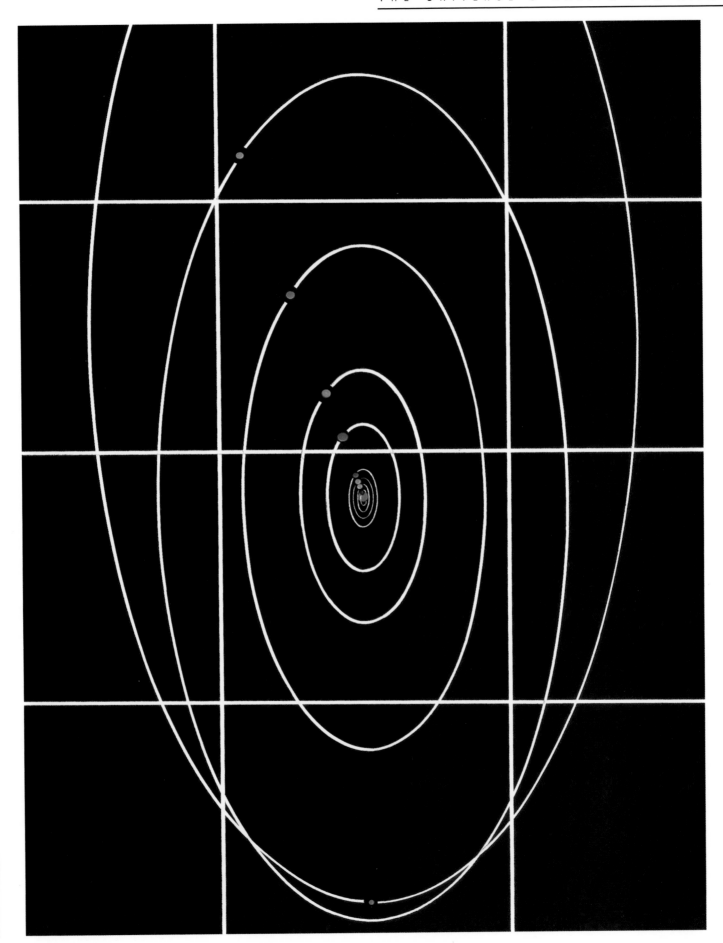

The Hubble Space Telescope, for instance – now that optical errors have been corrected – can examine likely stars for planets, as well as peer into the deepest recesses of space to research problems of the Universe's origins and final fate. In 1997, a

PIT STOP IN SPACE *Astronauts carry out a routine service on the Hubble Space Telescope while it is docked in the cargo bay of the space shuttle Endeavour.*

battery of sensitive equipment was turned on the comet Hale-Bopp, which flared across the sky in one of the most spectacular celestial events of the century. Analyses like these will help to solve the questions posed by the existence in space, and in cometary debris, of organic molecules – a possible origin of life on Earth.

If Earth's dominant life form has the curiosity, aggressive drive and technology to find out about its own origins and step off its planet into the inhospitable environment between the planets, then what might

the future hold? Is there any reason to suppose that human beings would hold back from further explorations, which would, in a geological eyeblink, take them not only beyond the confines of their own planet, but beyond their Solar System as well? For it has become apparent that the Earth's slow-moving processes have now produced something radically new – a species that can modify its home planet and leave it. Whatever happens to the Earth in the future, it has produced a dominant species with the potential to ensure its own immortality.

THE FIRE OF CREATION

1

FAMILY AFFAIR *William Herschel and his sister Caroline study the heavens.*

TWO OF THE MOST STARTLING AND CONTROVERSIAL CONCLUSIONS OF MODERN COSMOLOGY — THE STUDY OF THE HISTORY OF THE UNIVERSE — ARE THAT THE UNIVERSE HAD A DEFINITE STARTING POINT IN TIME, AND THAT AN UNDERSTANDING OF THE EARTH AND ITS LIFE FORMS DEMANDS AN UNDERSTANDING OF THE FIRST MOMENTS OF CREATION. THE RADIATION AND PARTICLES EMERGING FROM THAT PRIMORDIAL CATACLYSM WERE THE RAW MATERIALS FOR STARS, THE CRUCIBLES THAT COOKED UP ALL THE HEAVY ELEMENTS THAT WOULD ONE DAY BE DRAWN TOGETHER TO BECOME THE STUFF OF GALAXIES, SOLAR SYSTEMS AND PLANETS, AND — ON ONE PLANET AT LEAST — OF LIFE ITSELF.

NEIGHBOURS *This galaxy is a satellite of the Milky Way.*

PARTICLES TO STARS

The discovery that the Universe is constantly expanding suggested the idea that it had once been squashed together, and then exploded in what is now known as the Big Bang, creating the raw material of today's Universe.

The idea that the Universe had a beginning is not new. 'In the beginning, God created Heaven and Earth' is a sentence echoed in many religions, legends and myths. But because that attitude was obviously mythical and smacked of divine intervention, philosophers and scientists from Aristotle onwards have usually preferred to assume that the Universe is eternal. Only quite recently, since about 1970, has a new view been accepted: that events in the first second of the Universe laid down the conditions for the formation of the stars, our planet, life and us. The words 'In the beginning...' have a new, vital meaning.

In the 1920s, such a suggestion would not have been taken seriously. At that time, it was still fashionable to dismiss the notion of a beginning as mere myth. The Universe was the firmament of stars, fixed, infinite and stable, even if it was becoming clear that it was far larger than it had once appeared. Throughout the 19th century, ever more

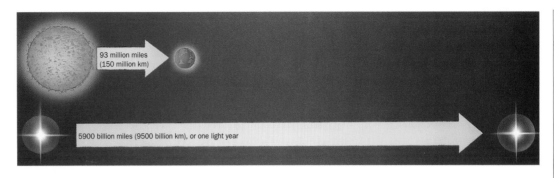

SPEED OF LIGHT *Light from the Sun takes about eight minutes to reach Earth. Light takes one year to travel between stars that are one light year apart.*

93 million miles
(150 million km)

5900 billion miles (9500 billion km), or one light year

powerful telescopes revealed that some of the misty patches known as nebulae (from the Latin word for 'clouds') were in fact whole systems of stars, so far away that they seemed a single, vague mass, as do sheep on a distant hillside. (Confusingly, some of the 2000 or so nebulae recorded in the 19th century turned out to be genuine clouds – diffuse clouds of dust.) In particular, the oblong patch in the constellation of Andromeda, visible to the naked eye, was now seen to be a great spiral of stars, at some vast but indeterminate distance.

At that time, distances between stars provided the measure of the Universe, and the speed of light was the best yardstick. Scientists knew that it took light, travelling at 186 000 miles (300 000 km) a second, eight minutes to travel to Earth from our Sun, several hours to cross our Solar System, four years to reach the star nearest to our Sun, and at least tens of thousands of years to reach the most distant stars.

Then in 1923, the great American astronomer Edwin Hubble, working with a new telescope on Mount Wilson, near Pasadena, discovered that the Andromeda spiral was an entity on its own, entirely separate from our own body of stars. The spiral, holding 100 billion stars, was 2 million light years away (one light year equals 5900 billion miles/9500 billion km, the distance travelled by light in a year). It was a sibling of our own Galaxy, the Milky Way. Suddenly, the scale of the Universe multiplied more than a thousandfold, for if Andromeda was another galaxy, so were countless other nebulae. And most of them would be very much farther away.

SIBLING GALAXY *The Great Spiral that appears in the Andromeda constellation lies 2 million light years beyond the Milky Way.*

For a few years, it was assumed that these galaxies were stable, a new kind of firmament to replace the stellar one. No one then suspected that these great island universes were merely atoms of a far greater reality, and that the whole of creation was in permanent flux.

In hindsight, scientists might have guessed there was something odd about prevailing assumptions, because they had long known of a paradox noted in 1826 by a German astronomer, Wilhelm Olbers (he was not the first to notice it, but he worked out the details). Olbers' Paradox can be understood in terms of a question. If the Universe is static and infinite, then every line of sight will eventually lead to a star (or galaxy – the argument remains the same). The more distant stars or galaxies are fainter, but even so there should be a solid glow of light across the whole night sky. So why is the sky dark at night?

AN EXPANDING UNIVERSE

The answer must be that the Universe is not static and/or infinite, which means that it must be dynamic and/or limited. However, no one saw the implications. Even Albert Einstein, whose theories of relativity had already redefined the nature of the Universe, refused to make the leap. His equations told him that the Universe was expanding, but he simply could not accept it, and in 1917 he inserted a 'cosmological constant' into his calculations in order to make the Universe static. He said later that this was the biggest blunder of his life.

The expansion, when Hubble conclusively revealed it, was a shock. The discovery came as Hubble analysed light from dozens of galaxies in a spectroscope. He noticed something curious: the light from them all

GALAXY MAN *Edwin Hubble discovered the existence of other galaxies beyond ours, and established that they are all moving away from each other.*

EINSTEIN'S ODD UNIVERSE

In defining his theories of how the Universe works, Albert Einstein virtually wrote the agenda for modern physics and astrophysics. His ideas – laid out in two theories of relativity, the Special Theory (1905) and the General Theory (1916) – link the world of everyday experience down to subatomic scales and up to cosmological ones. So far his theories have withstood every test.

The Special Theory is founded on the principle that the measured speed of light is constant, regardless of the speed of the light source or of the observer. In addition, the speed is the same for all observers, no matter what their motion relative to each other. The speed of light is the ultimate speed limit of the Universe – nothing can travel faster – and it is

CURVING SPACE A laser beamed into positively curved space might eventually arrive back at its starting point. Beamed into flat or negatively curved space, it would go on for ever.

connected with matter and energy.

At speeds approaching that of light – 186 000 miles (300 000 km) per second – strange things happen. Time begins to run more slowly, so material processes slow down; objects contract along their own length; and they become more massive. In Einstein's 'Twins Paradox', one fast-moving space-travelling twin would age more slowly than his stay-at-home brother.

Einstein also showed that matter and energy are interchangeable. His equation $E=mc^2$ states that the energy (E) of an object is equated with its mass (m) multiplied by the square of the speed of light (c). In effect, a small amount of matter can be converted into a huge amount of energy: 2 lb (1 kg) of matter contains enough energy to light 1 million homes for ten years.

These bizarre predictions have been confirmed by countless experiments. Orbiting atomic clocks, accurate to one part in many millions, have been measured

running slower than Earth-based clocks. The energy released from small lumps of uranium in nuclear explosions and power plants equates with Einsteinian predictions. And in the Big Bang theory, the whole universe of matter can be explained as deriving from a minute universe consisting entirely of energy.

The General Theory extends relativity to gravitational fields. Where Newton's laws are based on the 'commonsense' view that all objects are attracted to each other along straight lines, Einstein concluded that the presence of matter distorts space (with its three dimensions) and time, forming space-time, a four-dimensional matrix that is curved. The relationship between our commonsense world and Einstein's deeper reality is rather like the relationship between a map, which is flat (two-dimensional), and the Earth's surface, which is curved (three-dimensional). In our world, we experience a three-dimensional view of a four-dimensional reality.

The presence of matter creates local curvatures in space – as mountains do on the Earth's surface.

SPEEDING BY If a spaceship passes Earth at almost the speed of light, observed from Earth (above) the spaceship contracts and time there slows down. Seen from the spaceship (top), Earth contracts and time there slows down.

In another common comparison, a star is like a metal ball on a rubber sheet. It distorts the surface so that other balls nearby tend to roll towards it down the gravitational slope created by its existence.

Exactly how the Universe is curved no one yet knows. If it is positively curved, it may be finite in size yet have no boundary, like the surface of the Earth. If a spaceship could travel in an apparently straight line far enough, it would end up back where it started. Or it may be negatively curved, in which light follows an open path, never returning. The type of curvature depends on how much matter there is in the Universe, which remains one of the greatest unsolved problems of modern physics.

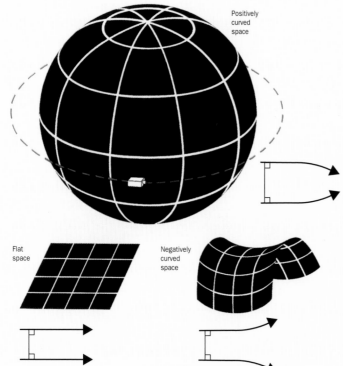

Positively curved space

Flat space

Negatively curved space

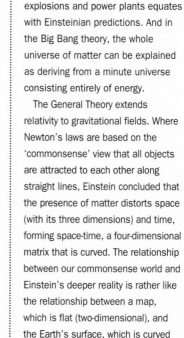

shifted in colour, and the only explanation was that the sources themselves were all moving. As the 19th-century Austrian physicist Christian Doppler had noted when he first described the effect, the result is similar to that of sound waves from a moving source.

Anyone who has heard an ambulance siren knows that the sound is higher-pitched as the siren approaches and grows lower-pitched as it moves away. This is because the sound waves are compressed as the sound approaches and stretched as it recedes. In the case of light, an approaching source compresses the vibrations or waves, which has the effect of shifting them towards the blue end of the spectrum; and a receding source stretches them, shifting them towards the red end of the spectrum.

Movement of some kind was not too surprising. Random movements would mean finding both red and blue shifts, but this is not what Hubble found. To his astonishment, almost all galaxies, except a few of the closest, were redder than they should be. In brief, they were all receding. Moreover, the 'redshifts' were proportional to distance. Those galaxies that were twice as far away were receding twice as fast. The actual speed, as defined by Hubble, rose by about 45 miles (75 km) per second for every 3.5 million light years – though the actual figures are much disputed, and in any case are temporary because the expansion is being slowed over time by the force of gravity.

At a stroke, Hubble had provided a yardstick for measuring the Universe out to the limits of visibility. Redshift, speed and distance all correlated to reveal the true scale of the Universe, with the result that 'redshift' has become a fundamental notion. The most distant galaxies are now generally accepted to be some 10 billion light years away, receding at over half the speed of light. This is to see the breakthrough in drastically simplified terms. Einstein's Universe is far stranger. According to his equations, the galaxies are not rushing away through space. Matter exists in a matrix of space. In a common metaphor, the galaxies are like dots on the surface of a balloon. It is space that is expanding, like the rubber of an expanding balloon, carrying the galactic dots ever farther apart.

This image can be used to explain another oddity. From where we sit in the Milky Way, it looks as if we are at the centre and all the galaxies are receding away from us. In fact, we are not in such a privileged position. The galaxies, like the dots on the balloon, are all receding from each other, with a

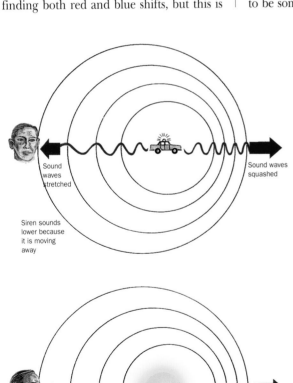

REDSHIFT *Just like sound waves from a receding siren (top left), light waves from a receding source are stretched, shifting them towards the red end of the spectrum (left).*

Sound waves stretched

Sound waves squashed

Siren sounds lower because it is moving away

Light waves stretched

Light waves squashed

Star looks redder because it is moving away

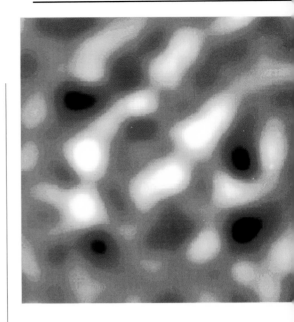

MICROWAVE RADIATION
The colours in this map of part of the sky are caused by temperature variations of 0.0001°C (0.00018°F). It is thought that these differences reveal the structure of the Universe 100 000 years after the Big Bang.

speed that is proportional to their distance apart. Any observer anywhere in the Universe will have the same impression.

WHEN DID IT BEGIN?

Once Hubble's observations were reconciled with Einstein's revised equations, the picture of a dynamic, expanding Universe was quickly accepted. But that opened the way to another breakthrough. For if the Universe was expanding, it must once have been smaller. Indeed, in theory it must once have occupied a very small space indeed. How long ago? Although astronomers constantly debate the figures, it is widely agreed that the Universe is between 13 and 20 billion years old – in round figures, about 15 billion.

This was an extraordinary conclusion, for it meant that in the past the uncounted millions of galaxies, each with their hundreds of billions of stars, were piled into each other in a single body of extreme density and mass, or existed in some other form as yet unguessed at. At a point in time, the body exploded into a fireball, producing the raw materials of the Universe.

For years, astronomers were loath to accept this suggestion. They believed that the

STEADY-STATER *Sir Fred Hoyle believed that new matter was constantly being created to fill an expanding Universe.*

idea of a beginning seemed nothing short of a return to a Biblical creation, one that flew in the face of scientific progress. For if, at the moment of creation, the Universe was born, so were the laws that govern it. By definition, there was nothing to be said about 'before' because there was no before. Science led, with a dreadful inevitability, to the end of scientific enquiry. The eminent British astronomer, Sir Fred Hoyle, was one of those who could not accept the implications. He derisively named the theory the 'Big Bang', and the name stuck.

For some 30 years, the debate continued. One body of opinion, rooted in the theories of George Gamow, refined the Big Bang model. Another line of thought, explored by Hoyle himself, suggested that the Universe was kept in a 'steady state' by the slow emergence of new material that formed new stars and new galaxies, filling the void left by those that were vanishing towards the 'edge' of the Universe.

There was a way forward, in theory and in practice. If the Big Bang theory was correct, then the initial fireball would have been accompanied by a stupendous burst of radiation. Actually, there would have been nothing but radiation, intensely hot, short-wave radiation. Since astronomers knew the speed of the expansion, they could predict

what would happen to that radiation. It would stretch, as the light from distant galaxies is stretched. Short waves would become long waves, and the temperature would drop, according to a simple formula: double the radius and the temperature falls by half. By now, the heat of the original cataclysm would have dissipated to something almost non-existent. In fact, in the background, behind and beyond later, more superficial events such as the evolution of stars and galaxies, there should exist a universal hum of long-wave radiation a few degrees above the temperature at which all motion ceases: $-273°C$, or $0K$, as it is known, after the 19th-century British scientist, Lord Kelvin.

As it happened, in the 1930s spectroscopic analysis of interstellar dust clouds had shown that their temperature was about $2.3K$. No one, however, thought to relate this to the temperature of the Universe as a whole. And by the early 1960s no one had backed the theory with new evidence.

The crucial discovery was made by chance. It happened that Bell Telephone Laboratories in the United States were experimenting with transglobal telecommunications systems, bouncing radio echoes from satellites. Since the satellites then did not amplify the signals, Bell needed a very sensitive receiver, and built what was called a horn antenna on Crawford Hill, near

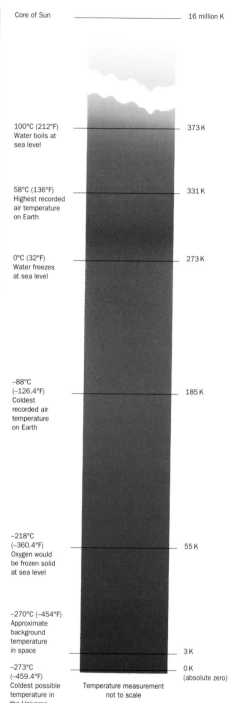

Core of Sun — 16 million K

100°C (212°F)
Water boils at
sea level — 373 K

58°C (136°F)
Highest recorded
air temperature
on Earth — 331 K

0°C (32°F)
Water freezes
at sea level — 273 K

−88°C
(−126.4°F)
Coldest
recorded air
temperature
on Earth — 185 K

−218°C
(−360.4°F)
Oxygen would
be frozen solid
at sea level — 55 K

−270°C (−454°F)
Approximate
background
temperature
in space — 3 K

−273°C
(−459.4°F)
Coldest possible
temperature in
the Universe — 0 K
(absolute zero)

Temperature measurement
not to scale

KELVIN SCALE *The Kelvin temperature scale (above) is named after the physicist Lord Kelvin (left). Born William Thompson, he defined absolute zero – 0 K – as the base line for measuring the temperature of the Universe.*

HEAT WAVES *Variations in temperature, shown by different colours, in the background radiation gave the Universe an irregular, rippled structure.*

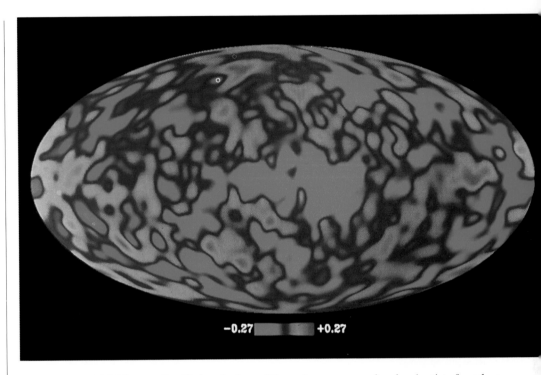

−0.27   +0.27

Holmdel, New Jersey. When the experiments were over, Bell handed the instrument over to two young radio astronomers, Arno Penzias and Robert Wilson. They were interested in measuring the background radiation from the Milky Way, but found a much stronger signal, coming not just from our Galaxy, but from everywhere, and the same strength in every direction. They were baffled.

In 1964, Penzias mentioned the mystery to an acquaintance at the Massachusetts Institute of Technology, who told him of an acquaintance of an acquaintance at Princeton, who had been involved in research into background microwave radiation. A few telephone calls brought the researchers together. Prediction and evidence tallied, and the following year the results were published. Later observations have confirmed and refined their discovery, that the Universe has a uniform background temperature of about 3 K (the accepted figure is now 2.7 K). Penzias and Wilson had unwittingly overheard the echo of the Big Bang, tapping radiation that dated from about 300 000 years after the beginning. From 1965, few doubted that the Big Bang theory was essentially correct.

For nearly 30 years, astronomers continued to refine the theory, exploring the ways

ELECTROMAGNETIC SPECTRUM
Visible light forms only a small part of the spectrum, which runs from the shortest-wavelength gamma rays to the longest-wavelength radio waves.

in which the initial burst of radiation had produced matter. But there was a problem. If the background radiation was uniform, the early Universe must have been uniform as well. But if there were no irregularities, where did galaxies and stars come from? Or if the early Universe was irregular, why was the background radiation so uniform?

Then in 1992 came the solution, from an unmanned satellite known as COBE (Cosmic Background Explorer). COBE, planned in the 1970s, was designed to test the Big Bang theory by seeing if the background radiation was as uniform as it appeared from Earth. In one sense, it was. COBE provided a super-fine refinement, measuring the universal temperature at 2.735 K. But COBE could measure far more accurately than that. It scanned the sky for a year at three different wavelengths, taking 70 million readings.

When these were analysed, scientists found that COBE had recorded minute differences in the background radiation: 30 millionths of a degree either way. Moreover, these variations were seen in all directions. Fortunately for Big Bang theorists, the early Universe had been both smooth and irregular, like an ocean ruffled by a breeze.

With this final piece of evidence in place, it became possible to tell the story of creation from the beginning and provide a convincing context for the emergence of the galaxies, the stars, our star, the Earth and us.

A MATTER OF GRAVITY

The story starts almost at the beginning, a split second after the Universe and everything it contained began its explosive expansion. Until recently, scientists had little *continued on page 22*

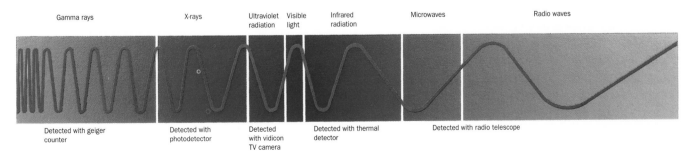

Gamma rays	X-rays	Ultraviolet radiation	Visible light	Infrared radiation	Microwaves	Radio waves

| Detected with geiger counter | Detected with photodetector | Detected with vidicon TV camera | Detected with thermal detector | Detected with radio telescope |

CAPTURING IMAGES FROM SPACE

In 1609 a Dutch optician, Hans Lippershey, patented a tube containing two lenses that made distant objects seem closer. It cost the equivalent of a pound or two, and started a revolution. It was a device like this that Galileo turned upon the Moon and planets and through which he saw the true structure of the Solar System.

Almost 400 years later, the US Space Shuttle *Discovery* placed a more sophisticated version of Lippershey's tube – the Hubble Space Telescope – in Earth's orbit. It, too, makes distant objects seem closer, at a cost of $1.5 billion, and enables scientists to see to the very ends of the known Universe.

Galileo's research inspired a mania for telescope building, in which the primary aim was to gather more light without losing detail. His telescope, with a lens about 1 in (2.5 cm) across, magnified about 1000 times. But simple lenses introduced distortions, for reasons analysed by Newton later in the 17th century. Newton got over the problem by devising a telescope that used a curved mirror, a principle that formed the basis of all light-based telescopes from then on.

Telescope sizes increased, different systems of mirrors allowed for better resolution (the clarity of the image), and differently

PIONEERS *Galileo's telescope (above left) had two lenses, which distorted colour. Newton's reflector (above right) gave greater clarity.*

shaped lenses overcame earlier distortions. One hundred years after Newton, Sir William Herschel built a 6½ in (16.5 cm) reflector, powerful enough for him to discover a new planet, Uranus. Inspired by this, he went on to build larger devices, culminating in a 48 in (122 cm) lens device set in a 40 ft (12 m) tube. Herschel used his telescope from 1789 to 1811, making dozens of major discoveries including planetary satellites, star-clouds and star-clusters.

In 1845 an Irish nobleman, William Parsons, the Third Earl of Rosse, built a massive telescope, 58 ft (17.5 m) long, with two mirrors each 72 in (183 cm) across. The device enabled Parsons to reveal the spiral structure of 'nebulae', later discovered to be separate galaxies.

Each increase in size brought new technological problems and solutions. Since all lenses and mirrors distort light, it is resolution, not magnification, that sets limits on the power of a telescope. Telescopes must track objects through the sky with absolute steadiness and without allowing gravity to distort the mirrors. In addition, the air imposes limits – peering into the depths of space is like looking at the world from the bottom of a pond through a surface ruffled by

GIANT EYE *Lord Rosse's telescope, of which this is a model, revealed that misty 'nebulae' were actually masses of individual stars.*

bad weather. Photography and spectroscopes have extended the effectiveness of telescopes, but the limits remain. Those limits were almost reached with the building of the 200 in (508 cm) Mount Palomar telescope in 1948.

At the same time, astronomy was undergoing another revolution. Light represents only a tiny fraction of the electromagnetic spectrum – 1 billion-million-millionth of its total spread. From ultrashort gamma rays with waves measured in billionths of a metre to radio waves hundreds of metres long, the Universe is a mass of unseen, very faint radiation. From the 1940s, astronomy acquired a new wing, radio astronomy, which taps into these whispers from space.

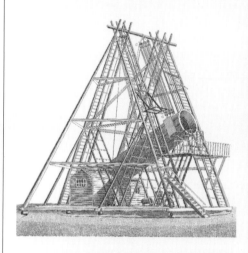

EXPLORING THE HEAVENS *Sir John Herschel's telescope enabled him to identify two new moons of Saturn and some 1500 galaxies.*

They can be intensified by building integrated arrays of receivers, which in effect creates a telescope hundreds of yards, even miles, across.

Today, astronomers rely on a range of Earth-based and satellite-borne devices that collect radiation at different wavelengths – visible light, radio waves, infrared and X-rays. The information gathered by these techniques is turned into digital form, then combined and processed into something vastly more detailed than anything dreamed of by the pioneers of modern astronomy.

TELESCOPES *Clockwise from top left: The Hubble Space Telescope; the Multiple Mirror Telescope, Mount Hopkins, Arizona; Britain's infrared telescope on Mauna Kea, Hawaii; Hale telescope at Mount Palomar, California; the Cambridge optical interferometer; the Very Large Array radio telescope, Socorro, New Mexico. Background: Stephan's Quartet group of galaxies.*

ORIGINAL THINKER *The Cambridge physicist Stephen Hawking revolutionised thinking about black holes.*

to say about what happened at the moment of the explosion, or before it, because it was only at this point that the laws governing our Universe came into existence. But over the past ten years the subject has become much debated in a bizarre line of speculation that owes much to the concept of 'black holes', the ultimate form of matter when crushed by gravity.

The idea that gravity forces matter into ever denser states – known as gravitational collapse – was first proposed in 1783 by John Michell in a paper to the Royal Society. It has been much refined this century, mainly from 1965 onwards by the Cambridge physicist Stephen Hawking, among others. The term 'black hole' was coined in the late 1960s by an American physicist, John Wheeler, and it was instantly seized on by scientists and the world at large as a key concept that offered a fundamental insight into the origin and fate of the Universe, as well as some of the objects in it. The concept forms the core of Hawking's book, *A Brief History of Time*.

In brief, a black hole is a body in which matter is compressed by gravity into a state so dense that nothing can escape from it,

not even light. A black hole can be any size, from a universe downwards: galaxy, star, planet, atom. In the extreme conditions in a black hole, any matter that is drawn into it is broken down to its simplest constituents, and squeezed into a 'singularity' of infinite mass and zero size in which the laws of physics, as we understand them, no longer exist. Any matter that falls into a black hole simply vanishes from view.

Some scientists, Hawking included, speculate that a black hole could be a 'wormhole' through which the matter is squirted, emerging as another universe with its own laws. Astronomers commonly describe the early Universe as a sort of black hole in reverse, with matter being forced outwards. Largely as the result of Hawking's work, it became possible to say something sensible about 'before', because the past and the future become one.

Meanwhile, we have to pick up the story a fraction of a second after the initial Big Bang, at a moment defined as 10^{-43} seconds. (Very large and very small numbers encountered in astronomy are often expressed using index notation, or powers of ten: 10^x denotes 1 followed by x zeros. For example, 10 000 is written 10^4. Very small numbers are expressed using a negative index: one ten-thousandth is written 10^{-4}.) Known as the Planck time, after the German quantum physicist Max Planck, 10^{-43} is the smallest unit of time to which scientists can ascribe any meaning. It takes us up to 1 millionth of a second after the explosion, when the temperature was 10 000 billion K. No matter can exist at such temperatures. The Universe was pure radiation.

It was during this stage, according to a theory substantiated by the COBE findings in 1992, that the expansion introduced minute but vital irregularities. As described by what is called inflation theory, in trillionths of that first second the Universe underwent a sudden inflationary burst, during

QUANTUM LEAPER *German physicist Max Planck originated the quantum theory of radiation and defined the smallest meaningful unit of time.*

which it expanded to perhaps 10^{50} times its original size. The effect of this expansion was both to iron out irregularities in the radiation, and at the same time to introduce irregularities into the earliest forms of matter – the basic particles that later joined up into atoms.

After 1 second, the temperature had dropped to 10 billion K, the temperature at the heart of an H-bomb explosion. Particles such as photons, electrons and neutrinos would have reacted with each other and with their mirror-image antiparticles, creating and destroying matter and annihilating each other in a subatomic maelstrom.

After about three minutes, the temperature would have dropped to about 1 billion K, which occurs today inside the hottest stars. At this heat, nuclear reactions are possible. Protons and neutrons paired off to produce the nuclei of deuterium atoms. In their turn, about 25 per cent of these nuclei each absorbed another proton and another neutron to produce helium nuclei. Leftover protons became the nuclei of hydrogen atoms.

The theory thus proposes that the very early Universe consisted of roughly 75 per cent hydrogen and 25 per cent helium. Since all the other heavier elements were formed later, and account for less than

HISTORY OF TIME *For a split second the Universe was pure energy. Then matter began to condense from the energy. After 3 minutes, simple atoms began forming. After 300 000 years nebulae were forming, and 3 billion years after the start, galaxies of stars began to form.*

3 MINUTES　　　　　　　　　　　　　　　　　　**300 000 YEARS**　　**3 BILLION YEARS**

1 per cent of the Universe's constituents, the theory also predicts the mix of elements in the Universe now. Fortunately for the theory, observation matches prediction, so scientists are fairly certain that the Big Bang theory is correct.

At the end of the period of nucleosynthesis, the cauldron of reactions settled. Expansion continued and the temperature dropped further, while the time-scale extended. Some 300 000 years after the beginning, the temperature had fallen to about 6000 K, as hot as the surface of the Sun. At this temperature, radiation and matter can no longer interact. The radiation was left to dissipate, declining ever more slowly to the 2.7 K measured by Penzias and Wilson in 1965, and later by COBE.

The matter, meanwhile, reacted in ways that became ever more complex. The hydrogen and helium that had emerged from the fireball were subject to two conflicting forces. One was the force of expansion, driving matter ever farther apart, stretching the gas ever more thinly. The other was the force of gravity, tending to drive molecules together, emphasising any lumpiness in the cloud of gas.

In a perfectly smooth universe, there would have been no gravitational collapse, because there would have been no lumps, nothing for gravity to work on. But the early Universe was not perfectly smooth. Its gas included those irregularities that were imposed by the inflationary period, and which were spotted by COBE. After about 1 million years, the denser parts of the gas – the lumps – began to coalesce into formless clouds. This lumpiness was minimal, with one atom for every 175 cu ft (5 m³) of space. But given time, gravity had its effect, and each cloud contained enough hydrogen and helium to form thousands of millions of stars in the coming aeons.

As the clouds became denser, forming rough ellipses, the slight internal movements imparted a spin, which increased as the clouds collapsed inwards under their own gravitational attraction, rather as a skater spins faster as she draws her arms in. Within the clouds, smaller-scale irregularities formed subclouds by the million, localised patches that also gradually became

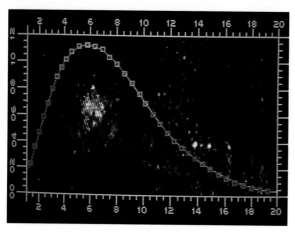

EVIDENCE FROM SPACE *Data from the COBE satellite shows that the Universe's brightness at ever-increasing wavelengths (squares) precisely matches the curve of an object radiating at a temperature of 2.7 K above absolute zero (red line).*

denser. It was from these small lumps that stars formed, eventually igniting to provide the Universe with the furnaces that would forge all the heavier elements and create the context for planets, and life itself.

As soon as the theory of the Big Bang was developed, scientists recognised the obvious long-term implications. If the Universe is expanding and if its speed of expansion is being slowed by gravity, then it should be possible to predict its future. They saw two possibilities: either it will go on expanding for ever, its stars and galaxies evolving, dying and dissipating in a scattering

of black holes, cinders of stars and diffuse dust; or gravity will overcome expansion, throwing the machinery into reverse, driving matter back together into what has come to be known as a Big Crunch.

To decide which is the more likely, astronomers need to know how much matter there is in the Universe. As yet, they don't. Current theories include the possibility that there is an invisible mass of 'dark matter'. Finding this, if it exists, is one of the principal tasks of modern astronomy. Meanwhile, it is clear that the Universe is expanding and that gravity counteracts this to some extent.

COSMIC COINCIDENCES

As this picture of the origin of the Universe emerged, so too did its relevance to human beings. Events that occurred 15 billion years ago defined laws that are universal: they work across the Universe, and for all time, defining the context within which we exist. Over the past few years, scientists have become fascinated by the narrow margins within which our Universe has developed. If any of those basic laws had been different, we would not be here to understand them.

Take one of the fundamental factors governing our Universe: the speed of expansion. As Stephen Hawking points out, if the rate of expansion one second after the Big Bang had been smaller by one part in 100 million billion, the Universe would long ago have recollapsed under the force of gravity. If the rate had been greater by such an infinitesimal amount, it would have overcome the force of gravity, forcing matter to dissipate so rapidly that no bodies as large as stars could ever have formed.

And what of another of the fundamental factors, the force of gravity – the mutual attraction between all particles of matter? If it were a little stronger, or weaker, the conclusions are similar: early collapse or rapid dissipation. By a set of seemingly extraordinary cosmic coincidences, the expansion started with just the right balance of energy (the

STELLAR FIELD *A section of our Galaxy shows a typical seedbed of stars and interstellar dust clouds. The bright object top left is another galaxy, Omega nebula M17.*

source of matter) and speed, and just the right amount of lumpiness, to provide the time and conditions needed for the evolution of complex objects such as stars and planets.

Many other cosmic coincidences have been crucial to the existence of the Universe as we know it, among them the other three fundamental forces that govern all matter: in order of decreasing strength, the 'strong nuclear force', which holds atomic nuclei together; electromagnetism, which binds atoms together (opposite charges of protons and electrons attract each other); and the 'weak nuclear force', which governs radioactive decay. How these forces came into being is the subject of research and debate, but one theory is that as the Universe cooled in the first few milliseconds of its existence, they were 'frozen' in their present forms, first gravity, then the strong force, then electromagnetism and the weak force together. It was this last event, the creation of the electromagnetic and weak forces, that could have released the energy to inflate the Universe and specify its combination of smoothness and lumpiness. From these initial conditions, all else flowed: galaxies, stars, elements, planets, life.

These 'constants' are so finely tuned as to spawn a whole body of thought based on the suggestion that the Universe is somehow created specifically to produce suns like our Sun, planets like our Earth and lifeforms like us. The argument picks up from the idea that the Big Bang was in effect a black hole spewing out radiation and creating matter, rather than sucking in matter and breaking it down into radiation. What if our Universe contracts again, leading in another few billion years to a Big Crunch? What if this forms a universal black hole, which in its turn spews out another universe? What if this is an endless process, in which each new universe emerges with its own set of laws, different on every occasion?

THE EVOLVING GALAXY

Each galaxy is formed by billions of stars, like the cells of a body, and provides the environment that nurtures them. And like the cells of a body, old stars are constantly dying and new ones forming in a continuous cycle of galactic regeneration.

The ancient Greeks explained the Milky Way in mythical terms. They related how the goddess Hera, sister-wife of Zeus and queen of the gods, spilled milk from her breast as she was suckling the infant Hercules. Theirs was a poetic response, but it forms the basis of modern terminology. The Greek for the Milky Way, *galaxias kyklos*, gives us our word 'galaxy'.

For 2500 years, science could not improve much on myth or poetry. Only in the 19th century did the two decouple. When William Wordsworth wrote that the daffodils he saw one afternoon in 1804 were:

Continuous as the stars that shine
And twinkle on the Milky Way,

he chose an image that was already becoming strained. Wordsworth would have been aware of research done over the previous 20 years by the astronomer, Sir William

Herschel, showing that the luminous veil spilling across the night sky was made up of stars in uncounted numbers – far more than the 3000 that are visible to the naked eye. 'Ten thousand saw I at a glance', Wordsworth guessed vaguely. But already science was ahead of him. Herschel knew he was dealing with stars in the hundreds of thousands.

And if Wordsworth had known the truth as revealed by modern cameras and telescopes, he might have chosen a different simile. A Milky Way of close-packed daffodils would cover 200 sq miles (520 km²) – half the Lake District; if they were as scattered as the stars are, each daffodil would

be 870 miles (1400 km) from its neighbour.

Better perhaps to return to the Greeks, for the milky band of starlight reveals something of our Galaxy's structure, which forms a great disc. The Milky Way is formed by the dense mass of stars leading out to the edge of the disc. The emptier, darker areas to either side open out onto the intergalactic wastes above and below the Milky Way's

GALAXIES GALORE *A vastly magnified pinprick view of deep space (right) reveals hundreds of galaxies which emerged, along with our Galaxy, from the universal seedbed. A side-on view of the Milky Way (below) shows it to be shaped like a flat disc with a thicker centre of brighter stars.*

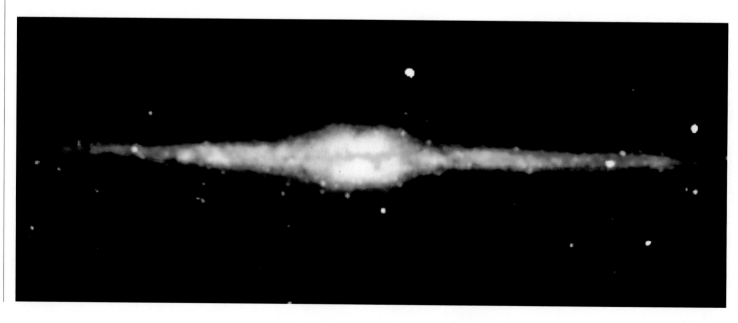

PATTERNS IN THE SKY

People like to see order in the stars by identifying patterns, such as the Plough, perhaps the most familiar constellation in the Northern Hemisphere, or the Southern Cross, the most familiar in the Southern Hemisphere. But there is nothing fixed about these pictures-in-the-sky. The Plough, the Southern Cross and all the constellations are merely join-the-dot shapes imposed as aids to remembering all the stars. Other cultures have seen other shapes in the same stars. To the Romans the Plough was the Great Bear, Ursa Major, still its formal name. To Americans, it is the Big Dipper.

It was the Roman habit – inherited from the Greeks – of relating star signs to legendary figures that gave us the familiar Latin nomenclature for the constellations. The ancient patterns were formalised in 48 constellations by the Graeco-Egyptian astronomer Ptolemy in the 2nd century AD. When the southern skies were first recorded in the 17th century, a German lawyer, Johann Bayer, made up a set of 12 new constellations. Later, additional constellations such as the Unicorn and the Giraffe emerged in the north, while the south acquired modern inventions such as the Air Pump and Telescope. Several suggestions – including the Balloon and the Printing Press – did not stand the test of time.

Traditionally, the constellations – especially the zodiacal signs – have astrological significance. But since the 17th century, scientists have known that the constellations are human interpretations; they have no astrophysical significance. Mostly, the stars have no connection with each other. Nor are they as fixed as they appear on the human time scale: 200 000 years ago there was no Plough, nor will there be in 200 000 years from now. Nor would they be valid for any alien life form – even from the nearest star, the Plough would have an extra star, our Sun.

Yet the constellations still have their uses. Sanctioned by tradition, they provide an international form of quick reference for astronomers. Stars are numbered or lettered according to their constellation: 61 Cygni (star 61 in the constellation Cygnus), Gamma Draconis. Our nearest neighbour is the Great Spiral in Andromeda. Pulsating stars are Cepheids, named after the first one recorded in Cepheus.

STAR SPOTTING *Against the background of stars a few constellations stand out, such as the Pleiades (the bright group right of centre).*

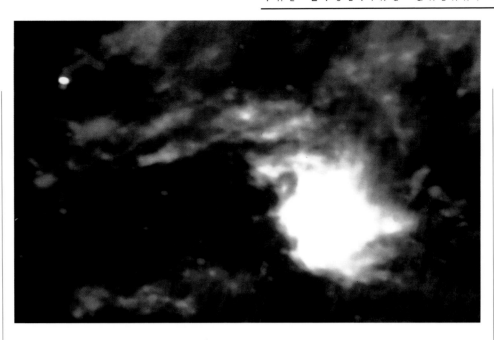

poles. It was Herschel who first realised the shape of the Galaxy, comparing it to a grindstone. He also suggested that our Sun was central, but in this he was mistaken. Much of the Galaxy is hidden behind clouds of dust, so he could only see parts of it. It was not until the 20th century that powerful telescopes and other instruments gathering infrared and ultraviolet radiation revealed the true proportions of our stellar surroundings.

Our Galaxy is an object on a prodigious scale. The great disc contains 100 billion stars – more than 20 stars for every person on Earth. It is 100 000 light years across, 25 000 times the distance between our Sun and the nearest star. This vast mass of stars is some 2000 light years thick, except at the centre, where it bulges out like the yolk of a fried egg, some 10 000 light years deep. The whole edifice is the shape of a catherine wheel with swirling arms of stars, and revolves once about every 250 million years. Our Sun, along with its planets, lies in a galactic suburb, about two-thirds of the way to the edge.

These figures are so far beyond normal concepts that it is easier to imagine them in more mundane terms. Daffodils are too large. The science writer and astrophysicist John Gribbin suggests a better working analogy. If you imagine the Sun as a Smartie (or M and M, if you are more familiar with American candies), then the nearest star would be another Smartie 90 miles (145 km) away. Even on this scale, the distances become literally astronomical, for the whole galaxy would consist of a group of Smarties three times the size of the Sun.

By convention, the canopy of stars seen from Earth are called the 'fixed stars', as opposed to the Moon and fast-moving planets, and they seem to be utterly apart, separated by 'space'. But the stars are not fixed at all, and they do not live apart from each other. If we could see them in fast motion, with one shot taken every 1 million years, they would appear in a whole variety of motions, swirling in orbit round the galactic centre, shifting restlessly under the influence of their neighbours' gravitational fields, and often buffeted by waves of gas

and radiation blasted from exploding stars. The galaxy is a dynamic structure, and its dynamism is vital to the formation of new stars such as our Sun.

CLUSTERS AND SUPERCLUSTERS

The formation of our Galaxy, like all the others, is tied closely to the formation of its individual cells, the stars, but it has a life of its own as well.

The embryonic galaxies, known as protogalaxies, tenuous clouds of hydrogen and helium, could begin to condense out of the remnants of the Big Bang after only 1 million years. At this stage, the Universe started to acquire a hint of its future structure.

GALACTIC FAMILIES

Galaxies are situated relatively close together – the space between them is about 100 times their diameters. Gravity groups galaxies into clusters, and clusters into superclusters. One supercluster of hundreds of galaxies, the so-called Great Wall, spans 500 million light years.

As the gas clouds collapsed in upon themselves under the force of gravity they acquired shapes, and usually a slow spin, with the larger clouds attracting and swallowing up smaller ones. As individual clouds formed, they remained bound by gravity to others, forming large family groups, which

STAR FOOD *A cloud of interstellar gas drifting past the Pleiades has been depleted (dark rectangle) as it fed the stars.*

we now see as clusters of galaxies. In fact, astronomers have recently realised that clusters of galaxies are themselves grouped into superclusters.

Large clouds became protogalaxies and within these, smaller clouds formed, broke up and merged again in a continuous jostle. Gradually, some areas became denser than others, a process that came to an end when the clouds began to trap heat. Each subcloud became a protostar, with an interior steadily heated as gravitational attraction caused matter to become more and more densely packed at the centre.

Eventually, in each star-cloud, when the temperature reached 10 million K, the heat caused ignition – thermonuclear fusion, the same source of energy that is released in the hydrogen bomb. In the centre of what had now become a star, the seething reaction prevented further gravitational collapse. In our Galaxy, as in millions of others, the original cloud had fragmented and refragmented to produce the first stars, by the million.

In round terms, astronomers estimate that there are 100 billion galaxies, each with 100 billion stars. Again, the numbers are stupefying and it may be easier to understand them by analogy. The proportional distances separating galaxies are very different from those separating stars. Imagine

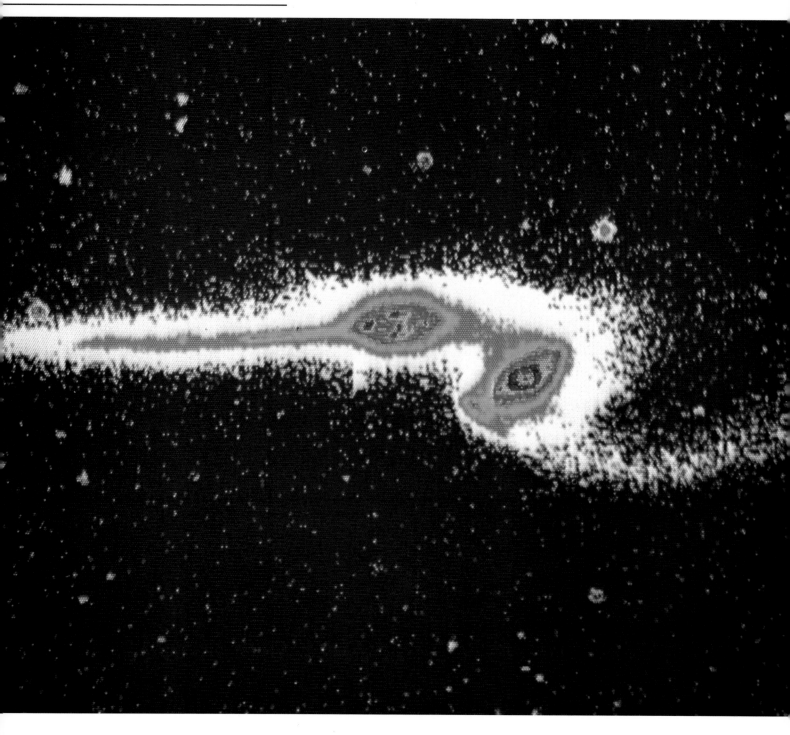

our whole Galaxy reduced to the size of a Smartie. The nearest neighbouring galaxy – the spiral in Andromeda – will be 5 in (13 cm) away, along with a small group of scattered 'chocolate crumbs'; together, these form our Local Group of galaxies. Other groups, almost 400 in all, each numbering dozens or even hundreds of galaxies, would be within about 20 ft (6 m), forming our Local Supercluster. In Smartie terms, the entire visible Universe would be about ½ mile (1 km) across.

SPIRALS, ELLIPSES AND RINGS

Our Galaxy is typical of millions of others, being a spiral galaxy with catherine-wheel-like arms. About 34 per cent of all galaxies are spirals, but there are other formats. Some galaxies (12 per cent) are elliptical or circular. About 50 per cent are irregular patches. A few are ring-shaped. All types come in a huge range of sizes, from dwarf ellipticals that are 10 million times smaller than the largest, to giant ellipticals up to 3 million light years across – so big that both the Milky Way and our nearest galactic neighbour, the Andromeda spiral, would fit inside.

With any luck, the process of galactic

evolution continues undisturbed for billions of years, as in the case of the Milky Way. In the Milky Way, the first stars formed the nucleus – the 'yolk' – of the Galaxy. Or rather, most of them did. A number of gaseous bubbles were left behind, outside the Galaxy, forming what are known as globular clusters, densely packed spherical clusters containing hundreds, thousands or even millions of individual stars. The nucleus of the Galaxy has been dubbed its geriatric ward, since it is populated by old stars. The galactic heart of the Milky Way contains a mystery – a powerful radio source far bigger than our Sun. Possibly it is a black hole, which would suggest that all galaxies may generate black holes at their centre.

SPIRALLING WORLDS

Ours is a spiral galaxy with arms circling a central nucleus (below left). Our Sun is located near the upper edge. Seen edge-on, the Galaxy is a thin disc, with a central bulge and a halo.
Below right: Spirals can be barred (left) or normal (right).

SUPERGIANT *Centaurus A is a large elliptical galaxy with a massively bright halo of billions of ageing stars.*

The real galactic action lies in the spiral arms, where much younger stars are constantly being born and dying in a cycle of life and death that provides the context for the birth of our own Sun. As the first stars settled into the nucleus, they left a vast disc of gas, mainly hydrogen. This was incredibly tenuous, far more rarefied than the most extreme vacuum on Earth, but it was still enough, when it broke up, to condense into stars by the billion. Clouds, however, only break up and condense when they are disturbed, which happens for three reasons.

Firstly, galaxies are subject to tidal forces from their neighbours. They do not have to be colliding for this to happen. Some, known as starburst galaxies, circle each other closely enough to stir up violent tidal effects that trigger a firework display of star-formation. In the case of our Galaxy,

the influence comes from the Andromeda spiral and the other smaller galaxies in our Local Group – about 30, mostly small ellipticals, and two fuzzy bodies visible to the naked eye, small satellite galaxies known as the Large and Small Magellanic Clouds. The Andromeda spiral also has its acolytes, seven dwarf ellipticals.

However much the Milky Way aspired to stability, its swirls of stars and dust would always have an oceanic surge forced upon

MEASURING THE DISTANCE TO THE STARS

To the naked eye, stars are enigmatic dots an unknown distance away from Earth, but some are closer than others. By changing position, an observer can detect their apparent movement against the stellar background. You can create the same effect by holding up a thumb and looking at it first with one eye closed, then the other. Your thumb changes its position against the background, an illusion known as parallax. Imagining a line from eye to eye as the base of a triangle, you can work out the distance of your thumb.

But even the nearest stars are so far away that the Earth is not large enough to act as a base for such a measurement. The largest platform available is provided by the Earth's orbit round the Sun. To measure a star's position twice, six months apart, is to measure from a base almost 200 million miles (320 million km) across.

This is easier said than done. For the early scientists the first task was to choose a close star, and the only way to do this was to select one that, on the basis of long-term records, seemed to be moving. In 1838, the German astronomer Friedrich Bessel used this technique to measure the parallax shift of a star known as 61 Cygni (the 61st in the constellation of Cygnus). It demanded phenomenal accuracy. The star's shift against its background was 0.3 seconds – less than $1/1000$ of a degree, the equivalent of the diameter of a coat button viewed from a distance of 10 miles (16 km). By simple geometry, he worked out that the star was 60 000 billion miles (96 000 billion km) away.

Since then, the parallax of thousands of stars has been recorded. But the farther away the stars are, the less they show parallax shifts, because they fuse with the background. Measurement of distance picks up where parallax stops with a variety of techniques. One of these involves analysing the motion of stars and the clusters to which they belong. Another involves analysing the spectrum, for the type of light reveals the star-type and its real brightness. A comparison between this and its apparent brightness provides a yardstick for calculating its distance.

Another method of measuring distance involves pulsating stars, a special group whose surfaces rise and fall, releasing and rebuilding energy rather like the lid of a boiling kettle. Some, known as Cepheids, change regularly, with periods of about five days that match their brightness. By finding Cepheids in

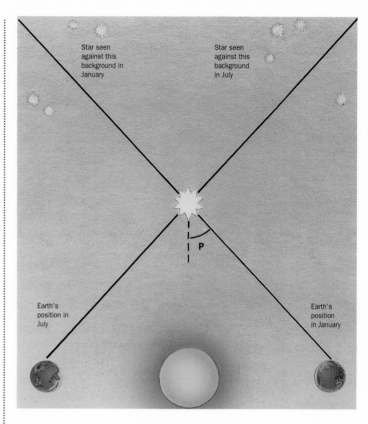

other galaxies and comparing their absolute brightness with their apparent brightness, astronomers can fix their distance.

For even greater distance, astronomers rely on light from whole galaxies. As all galaxies (except the very closest) are receding, their light is shifting towards the red end of the spectrum. Redshift correlates with speed, and speed with distance.

PARALLAX *By recording the shift in a star's position against the background six months apart, astronomers can work out the angle P and estimate the star's distance from Earth.*

By combining these techniques, astronomers can get the measure of the Universe, from the closest stars to the most distant galaxies.

---

them by its neighbours. That effect, though, is ironed out by the much more massive instabilities set up within the Galaxy itself. Though apparently frozen when viewed on a human time scale, the catherine wheel arms are actually in violent motion. They look as if they are made of stars swirling out from the galactic hub, but they are not. If they were, the stars would circle the hub rather like cream in a newly stirred cup of coffee, with dozens of circles corresponding

to the number of revolutions of the galaxy. In fact, they make only one or two turns round the centre, because they are formed by an entirely different process – by pressure waves, as separate from the matter in them as ocean waves are from the molecules of water they disturb. These waves, which move round the Galaxy at about 20 miles (30 km) per second, jolt the dust and gas and stars, pushing the material into the curving waves that we see as the Galaxy's

arms, and initiating the process of gravitational collapse that leads to yet more star formation.

The third cause of galactic disturbance is the stars themselves, which burn helium and hydrogen in nuclear reactions, releasing energy as they do so. When this cycle has run its course, larger stars explode into 'supernovae' and blast out elements such as uranium, copper, gold and silver into interstellar space. These enriched clouds collapse again into second-generation stars, such as our Sun.

All this activity makes even staid spiral galaxies seem a most unstable place. And they are – but oddly, these galaxies, our own included, should be a lot more unstable. Computer simulations show that the spiral pattern should not endure anywhere

A LEGEND IN THE SKY

According to legend, Queen Cassiopeia boasted that her daughter, Andromeda, was prettier than Neptune's daughters, the Nereids. Neptune sent a sea-monster, Cetus, to ravage the kingdom. An oracle said that to get rid of this scourge, Andromeda should be chained to a rock so Cetus could eat her, but she was saved by Perseus. All the characters have constellations or stars named after them.

near as long as it does. Something is acting to damp down the instabilities introduced by the gravitational effects of other galaxies, the individual galaxy's spin, the pressure wave and stellar explosions. It seems there is a lot more gas – in the form of so-called dark matter – than can be seen. Astronomers guess it is there as a halo ten times the mass of all the stars, but made up of a different kind of matter – a kind that does not emit light and is therefore invisible to our light-sensitive eyes.

Increasingly, astronomers have come to see this whole galactic process in biological terms, for it is reminiscent of the way in which living systems sustain their unstable pattern. Each cell in the human body lasts only for a few weeks or at most years, yet the system as a whole endures for a good three-score years and ten or even longer. In the same way, the galaxy's body is self-regulating, repairing itself over its lifetime by the death and rebirth of its cells, the stars.

RADIO SOURCES

Galaxies are dynamic structures, but until the middle of the 20th century astronomers tended to see them as rather peaceful objects moving in a gentle slow motion. It was the development of radio astronomy

in the 1950s that revolutionised this view.

Some galaxies are self-destructive. These, named after the astronomer Carl Seyfert, have a central region of hot gas in violent motion, with a nucleus of millions of suns crowded together. Unless galaxies have a fair amount of spin to force their stars apart, they can collapse, with results that were discovered only in the early 1960s. At that time, when radio astronomy was beginning to enrich our picture of the Universe, astronomers knew that there were many powerful sources of radio energy. But as few of these sources could be related to visible objects, no one knew what caused them or how far away they were.

In 1962, astronomers were given an opportunity to fix on one of these sources. It so happened that the Moon passed right in front of it. By timing the moment at which the Moon cut off the radio waves and the moment at which the signal reappeared, they were able to pinpoint the source. It was a star-like object labelled 3C 273 – star number 273 in the 3rd Cambridge Catalogue of radio sources (the catalogue, abbreviated as 3C, is based on surveys made at Cambridge, England, in the 1950s of galactic and intergalactic radio sources). However, when 3C 273's redshift (the amount of radiation emitted as it receded) was measured, it turned out to be far more distant than any star. It was small enough to look like a star, yet it was as bright and energetic as a galaxy, and at a galactic distance. In fact, if the relationship between redshift, distance and speed held true, it was 3 billion light years away and still receding at a good proportion of the speed of light.

Soon, other such objects were discovered, some with even more energy than the first. They were dubbed quasi-stellar sources, or quasars for short. Astronomers now know of around 1500 quasars. They are important in cosmology because they are, on the

whole, very distant and energetic. One of them is apparently receding at almost 90 per cent of the speed of light. The light that we can see coming from this quasar must have set out not long after the Big Bang, which means that we are observing a sort of astronomical fossil. Quasars are also important because their implications are still controversial: some scientists wonder if the evidence for these extraordinary objects might not have another explanation as yet unknown.

There are a whole range of other types of active galaxies that are powerful sources

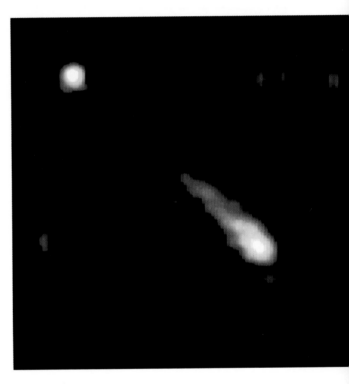

MAKING WAVES *In this image of quasar 3C 273, the colours indicate the intensity of its radio energy emissions, ranging from white (highest) to red (lowest).*

of radio energy and X-rays. Some are in the process of exploding. Quasar 3C 236 consists of two lobes that are 20 million light years apart – ten times the distance between our Galaxy and the Andromeda spiral – as the result of an explosion that took place over 10 million years ago. Another example is the elliptical galaxy M-87 (M stands for

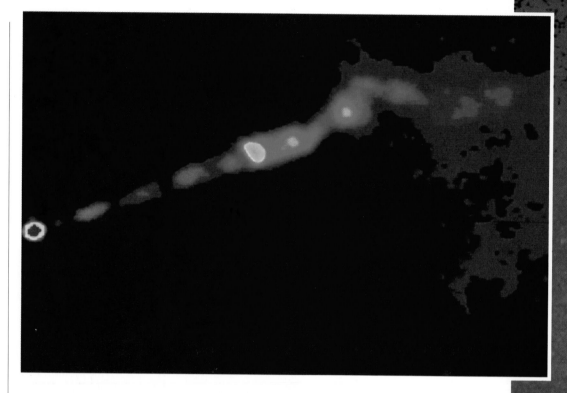

Messier number, after Charles Messier, 1730-1817, the first astronomer to catalogue star clusters and nebulae). M-87 has such a violent core that it is firing out a jet of stars and gas some 6000 light years long, ending in a blob that looks like a mini-galaxy in the process of formation. Other quasars simply seem to be massive sources of radio energy, like Cygnus A, a radio source 10 million times more powerful than our own.

There is little doubt that this galactic activity is the result of explosions in the centre of the galaxy. The ultimate source of power in each case is probably a massive black hole. The black hole sits like a spider in the centre of a web, drawing material into its massive gravitational field. Since the black hole is spinning, the material spirals into a disc shape, like water going down a plughole, being compressed as it falls inwards. It cannot all fit in at once and some material erupts, causing a surge of gas and emitting a beam of radio energy. The galactic core becomes, in effect, a dynamo producing radio waves.

For us on Earth, such events may seem too remote in time and space to have anything but intellectual relevance. However,

GALACTIC ENERGY *The galaxy M-87 fires out a jet of material some 6000 light years long. A false-colour image derived from radio emissions (above) shows the shape of the jet, while an optical image (right) relates it to the galaxy's stars.*

they are of particular significance because they underline the narrow confines within which complex systems – from solar systems to individual life forms – develop.

On the galactic scale, the Universe is pretty crowded. Since galaxies are relatively close together, bound by gravity into clusters and superclusters, they exert a tidal pull on one another. Some, wheeling in a dance extending over billions of years, are so close that they collide, pulling great streamers of stars from each other. Indeed, evidence from the Hubble Space Telescope suggests that all elliptical galaxies are the product of two spirals colliding. Only more sedate and independent objects, such as spiral galaxies well away from their neighbours, are likely to have the necessary balance between stability and activity that creates the conditions for complex objects such as humans to exist.

COOKING UP THE ELEMENTS

Stars are the crucibles that transform the simple materials left over from the Big Bang – in the form of clouds of gas and dust – into the multitude of more complex elements from which the planets are eventually formed.

The galaxy has a life of its own, but that life derives from its individual cells, the stars. As well as driving the large-scale evolution of the galaxy, their fires also power the evolution of smaller-scale objects, processing the raw material from which new stars, along with their planets, are made.

When the first stars emerged from the primordial gas, they formed in a range of sizes and temperatures, from hot, bright ones to cool, dim ones, but they all started off in the same way, as furnaces of compressed hydrogen and helium. Only in later life did their different sizes dictate different histories. Then, as aeon succeeded aeon, and stellar generation succeeded generation, stars began to change, enriching their surroundings and themselves in a self-driven evolutionary process that eventually, after about 10 billion years, produced stars like our Sun. To understand the special features of our Sun, we need an understanding of how other types of star evolve, and what they contribute to the process of creation.

Since stellar formation is triggered by motion, and the galactic arms are the most volatile regions, it is in the arms that younger stars predominate. There, a whole range of new stars continue to form, at the rate of about ten a year, from dim, cool, red ones one-tenth the mass of the Sun to hot, bright ones 50 times its mass. This is not a simple process, because as the clouds condense they begin to pick up rotational speed, and often break up under their own centrifugal force.

THE LIFE STORY OF A STAR

The life of any star is dictated almost entirely by its size. In any new star, hydrogen burns to helium, releasing energy that makes the star shine and supports its outer shell against the force of gravity. Medium-sized stars, such as our Sun, are relatively stable, burning their hydrogen fuel for something like 10 billion years, but larger

GAS CLOUD *The Crab Nebula is a cloud of gas blown from a star that exploded into a supernova, replenishing the Universe with the raw ingredients for new stars.*

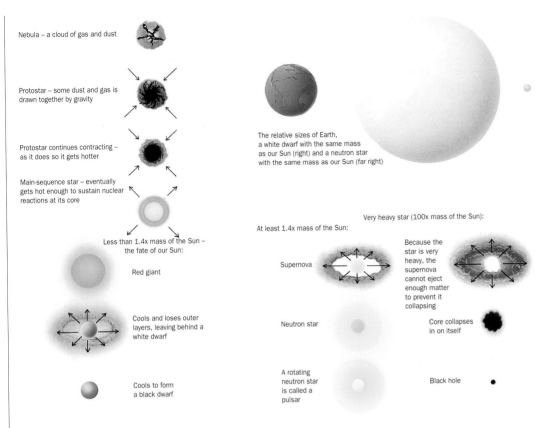

Nebula – a cloud of gas and dust

Protostar – some dust and gas is drawn together by gravity

Protostar continues contracting – as it does so it gets hotter

Main-sequence star – eventually gets hot enough to sustain nuclear reactions at its core

Less than 1.4x mass of the Sun – the fate of our Sun:

Red giant

Cools and loses outer layers, leaving behind a white dwarf

Cools to form a black dwarf

The relative sizes of Earth, a white dwarf with the same mass as our Sun (right) and a neutron star with the same mass as our Sun (far right)

Very heavy star (100x mass of the Sun):

At least 1.4x mass of the Sun:

Supernova

Because the star is very heavy, the supernova cannot eject enough matter to prevent it collapsing

Neutron star

Core collapses in on itself

A rotating neutron star is called a pulsar

Black hole

THE LIFE AND DEATH OF STARS
All stars form in the same way, but their fate depends on their size. Whereas our Sun will expand into a red giant and then cool and contract, larger stars form supernovae. Very heavy stars explode and then collapse to the point where they form a black hole.

gy, and ever more complex nuclear reactions. Helium burns to carbon and oxygen, which burn to silicon, which burns to iron. At each stage, other rare elements are also produced: neon, magnesium, sulphur, phosphorus. When the core consists of surrounding shells of silicon, carbon, oxygen, helium and hydrogen, the end is near. Indeed, it has approached ever more rapidly as the processes succeeded each other, for the bigger the star, the faster it burns.

In a heavy star, say 25 times the mass of the Sun, hydrogen burns for some 7 million years, helium for 500 000 years, carbon for 600 years, neon for 1 year and oxygen for six months, while the conversion from silicon to iron takes but a single day. Once all the nuclear fuel is used

stars are more profligate. A star ten times the mass of the Sun may be 1000 times brighter, consuming its hydrogen fuel 100 times faster.

In stars about the size of our Sun, there comes a time after about 10 billion years when all the hydrogen is exhausted, leaving a core of helium 'ashes'. The nuclear furnace starts to die, allowing gravity to resume its work. At a certain point the core of the star collapses, increases in temperature and re-ignites, this time burning the less volatile helium to form carbon. The new burst of activity pumps energy into the mantle, driving it outwards. The star balloons up, becoming a red giant (like Betelgeuse, which is so huge that it would engulf Mars if it were placed at the centre of the Solar System). Our Sun itself will become a red giant about 5 billion years hence. Eventually, the outer canopy lifts completely clear of the star beneath and drifts away into space, making a beautiful, three-dimensional smoke-ring known as a planetary nebula.

What remains is a superdense clinker, a white dwarf. These little stellar remnants, which have the same mass, or quantity of matter, as our Sun but are only the size of the Earth, and are so dense that a matchbox-sized amount of them would weigh many tons. But as long as they are less than a critical size – 1.4 times the mass of the Sun – they simply cool off for ever, becoming a dead stump of a star, a black dwarf.

Stars with a greater mass have a more interesting destiny, with wider consequences for the rest of the galaxy and the life forms in it. They are bulky enough for gravitational collapse to generate ever higher temperatures, ever greater ener-

SUPERNOVA REMAINS *An X-ray image of hot gas reveals the remnants of a supernova explosion that was recorded in China in AD 185.*

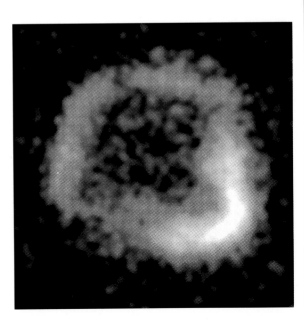

CLOUDS BETWEEN THE STARS

Between the stars lie regions of gas and dust that are both products of long-dead stars and the raw material for new ones. Some nebulae are the outer covers of red giant stars that have cast off a smoke-ring of gas. These planetary nebulae, of which some 1000 are known, disperse over 10 000 years. Sometimes, two stars interact and throw off tidal waves of gas – so-called novae. Occasionally, massive stars explode into 'supernovae', ejecting gas rich in new-formed elements and travelling at speeds of some 6000 mph (9600 km/h). These clouds, swirling slowly as they disperse, gradually collapse to form new stars.

BUBBLING WITH LIFE
The Orion Nebula is a bubble of gas in the constellation of Orion inside which new stars are being formed.

up, the result is cataclysmic. The iron core – which has a mass about the size of the Earth – implodes under the force of gravity, which makes the iron atoms break down into neutrons.

In fractions of a second the outer shells collapse inwards, reaching speeds of 25 per cent of light speed. They smash into the new neutron core and explode outwards again. In the process, a variety of elements heavier than iron are forged – lead, gold, uranium and others. Almost instantaneously, some 90 per cent of the star is blasted into interstellar space. The sudden release of energy multiplies the star's brightness many thousandfold, making it anything up to 10 billion times as bright as our Sun and producing what seems to be a new star in the night sky.

These events were named 'supernovae', by analogy with other, lesser stellar explosions known as 'novae' (sometimes known as novas). In fact, novas are caused by a completely different chain of events involving two gravitationally bound stars, one of which loses matter to the other. In a supernova explosion, the whole star may be blown to bits. In any event, in the explosion a star returns its processed fuel to the Universe, much enriched.

If the supernova remnant is more than the critical 1.4 solar mass, it collapses further to compress its matter so densely that its atomic structure breaks down completely, forming one of two exotic, high-density objects: a neutron star or, if it is bulky enough, a black hole. These things may seem remote from the biography of the Earth, but through their gravitational influence, they all contribute to the violent fluctuations in the interstellar gas and dust that are crucial to the development of stars in the first place.

FOSSIL POWER

Nuclear power is produced from uranium, one of the elements produced in supernova explosions and blasted into space to become the raw material for new stars and planets. When you use electricity generated by nuclear power, therefore, you are turning on your light using the fossil remains of a massive star.

NEUTRON STARS AND BLACK HOLES

A neutron star, in which electrons and protons are forced together to form neutrons, may be as little as 10 miles (16 km) across. It is an extraordinary object, with gravity perhaps 100 billion times as powerful as the Earth's. So intense is the gravitational field that an object would have to attain about 80 per cent of the speed of light to escape from it. Though decidedly odd, neutron stars are simple objects whose structure is well understood. They have a fluid core, two types of crust and a surface about 1/2 mile (0.8 km) thick, which is so smooth that its 'mountains' will be only a few inches

NEW STARS FROM OLD
Molecular hydrogen and dust form gas pillars about one light year high that are seedbeds of new stars.

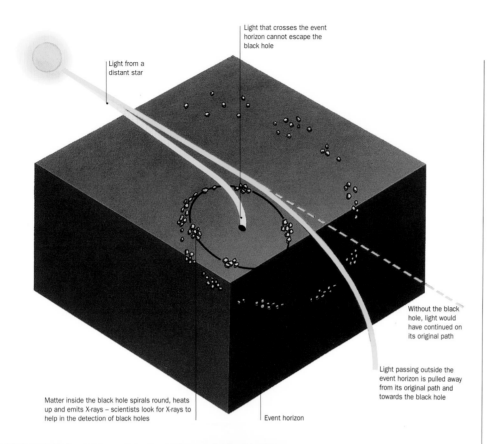

Light from a distant star

Light that crosses the event horizon cannot escape the black hole

Without the black hole, light would have continued on its original path

Light passing outside the event horizon is pulled away from its original path and towards the black hole

Matter inside the black hole spirals round, heats up and emits X-rays – scientists look for X-rays to help in the detection of black holes

Event horizon

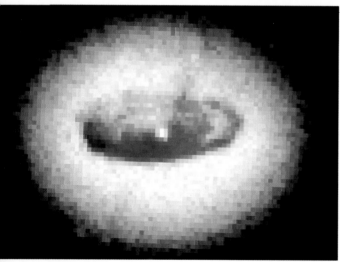

NO ESCAPE The gravitational pull of a black hole is so strong that space-time curves towards it. Past a certain point, the event horizon, nothing can escape being pulled in.
Left: At the centre of the galaxy NGC 4261 is a disc of dust and gas being sucked in by a black hole some 1.2 billion times the mass of the Sun.

and were named the Crab Nebula after the crab-like sketch made of it by the British astronomer Lord Rosse. It is now known that the original star, some 6300 light years from Earth, would have been invisible until it turned into a supernova, making it briefly 200 million times as bright as the Sun.

Another reason for the Crab's fame was that it is one of the most powerful sources of radio energy. No one knew why until 1967, when a star right at the centre of the Crab was seen to be behaving in a most peculiar way. It was pulsing on and off 30 times a second, something so extraordinary and so apparently inexplicable that its discoverers, Jocelyn Bell and Antony Hewish at Cambridge University, suggested it might be a beacon set up by an alien civilisation. For a while, astronomers referred to the star as LGM 1 ('Little Green Men'). In fact, this was the first recorded 'pulsar', the neutron star remnant of the supernova explosion, which in collapsing had retained all its former angular momentum, or rotation, and was now spinning on its axis 30 times a second.

Since then, some 350 pulsars have been identified, one of which is revolving at 642 times a second, with a regularity that makes it more accurate than an atomic clock. Actually, it loses just 300 millionths of a second every 1000 years, which is not simply an indication of its strangeness, but also of the astonishing accuracy with which astronomers can measure its signals. As a result, astronomers know that some pulsars have a slight irregularity. The only explanation at present is that they have near neighbours in orbit around them – in other words, they have one or more planets.

A neutron star is not the ultimate form of matter, however. If a neutron star is more than about three solar masses, the neutrons will not be enough to withstand further collapse under the force of gravity. All the matter showers inwards towards a central point, squashed into a mass of zero volume and infinite density, but with a gravitational field proportional to its mass. The matter forms a black hole.

In theory, a black hole can be of any mass, even though it has no size. Once matter is compressed into a small enough

high. They also have an 'atmosphere' a few yards thick of atoms, electrons and protons.

Moreover, all its previous volume has now contracted hundreds of thousands of times, accelerating its rotational speed along the way. The consequences of this only became apparent in 1967, when astronomers discovered something odd in the middle of a cloud of gas known as the Crab Nebula. The Crab is one of the most famous objects in astronomy, partly because

it is the remains of a recorded supernova. On July 4, 1054, Chinese astronomers noted the sudden appearance of what they called a 'guest star', as bright as Venus. 'It had pointed rays on all sides,' they recorded, 'and its colour was reddish-white.' It dominated the night sky for weeks, and would have been noticed by any society with any knowledge of the stars. In Chaco Canyon, New Mexico, there is an Indian rock painting that may well record the event, though as yet no references to it have been found in Europe.

The star's remnants were discovered in the 18th century as a bright, fuzzy cloud,

FINGER ON THE PULSE *British radio astronomer Antony Hewish was co-discoverer of the first known pulsar, a spinning stellar remnant.*

volume – a critical size known as the Schwarzschild radius (after the German astrophysicist Karl Schwarzschild) – it becomes a black hole. The Earth would become a black hole if it were to be compressed into a globe less than 1/2 in (1.3 cm) across. The Sun's Schwarzschild radius is 1³/₄ miles (3 km). It is possible that galaxies

have massive and powerful black holes at their centres, and that whole galaxies could form black holes. Indeed, it is possible to conceive of the whole Universe collapsing into, and emerging from, a black hole.

Though only stars of three solar masses or more will collapse into black holes under the force of gravity, it is possible (again in theory) for a supernova to blast matter inwards so powerfully that the remnant is a minute black hole. This means that the granddaddy of all supernova explosions – the Big Bang – might have created

unknown billions of miniature black holes, objects with the mass of mountains but the size of subatomic particles.

These strange objects have seized the popular imagination because they have such peculiar properties. Once within its gravitational clutches – the so-called 'event horizon' defined by the Schwarzschild radius – nothing can ever escape, not even light itself. However much matter is sucked in, it all falls into the central point of infinite density and zero size – the point that is known as a 'singularity'.

Ever since the idea of black holes was proposed in the 1960s, they have been treated as real, though in a sense they remain theoretical things. They cannot be seen. Nor can their absence be seen – there is no 'black hole' in space where a

---

THE MESSAGE OF LIGHT

Visible light provides the first vital clues to the nature of stars. Ordinary light consists of a range of colours, which can be seen when light is split up in a rainbow. The spread of colours is called a spectrum. When light passes through anything that bends the rays (whether a glass prism or a raindrop), the light separates according to wavelength. White light contains all the visible wavelengths, and the spectrum forms a continuous band from violet to red.

In the early 19th century, the German physicist Joseph von Fraunhofer discovered many dark lines – he recorded 576 – in the spectrum of light from the Sun. Subsequent research showed that any atom absorbs light of a particular colour. If light is shone through a gas, the elements making up that gas absorb their own colour, and that wavelength cannot contribute to the spectrum that emerges. The spectrum shows a blank spot – a dark line, like a shadow of the missing radiation.

In a star, the outer layers contain many elements, and these absorb particular wavelengths of the star's light. When the light is spread out into a spectrum, Fraunhofer lines appear; some 25 000 of them are now known, all associated with their own chemicals. Thus every star can be fingerprinted, showing the chemicals present in its outer regions. In addition, analysis of the strength of the lines in a star's spectrum reveals the temperature of the material producing the lines.

Once stars are fingerprinted, it is possible to see what other shadows exist on the spectrum, and deduce that the missing radiation has been absorbed by gas and dust lying between the stars and Earth. This gives clues to the nature of interstellar clouds.

Since the lines relate to particular positions on the spectrum, and since a fast-moving source 'stretches' the spectrum, lines that are shunted to left or right of their true position indicate motion. Typically, galaxies

are receding, so the lines are shifted towards the red end of the spectrum by an amount that defines the speed of recession, or redshift.

FINGERPRINT *The missing wavelengths in light from a star can be analysed to find out what the star is made of.*

The outer layers of a star, or a gas cloud, absorb some of the spectrum of light

Absorption lines caused by missing wavelengths

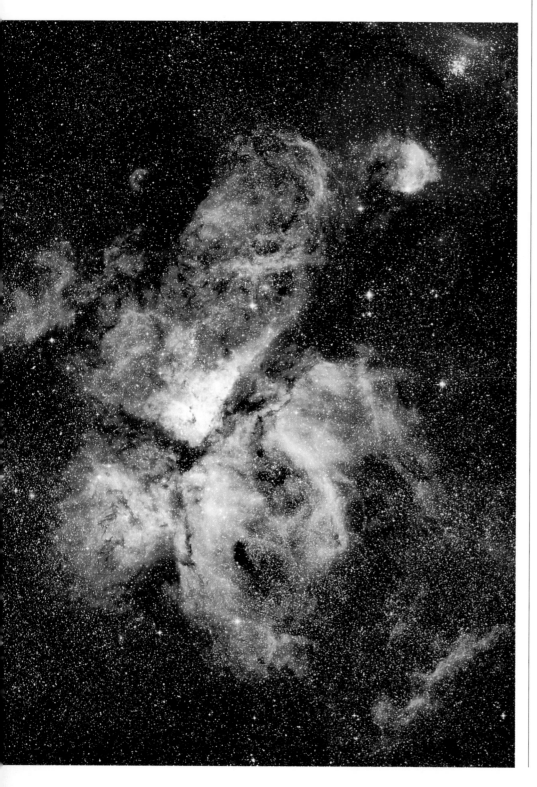

star might be. If a black hole could exist in isolation, light from behind it would bend around it, so the sky would look undisturbed. Its presence would have to be inferred, as matter swept into it accelerates to light speed, heats up, and emits powerful radiation in the form of X-rays.

In fact, as Stephen Hawking has shown, black holes are not as black as they were once painted. They leak radiation, very slowly, smaller ones leaking faster than large ones. A black hole with the mass of the Sun would evaporate in 10^{60} years, which is about 50 times as long as the Universe has already existed. Any primordial black holes formed in the Big Bang should be in the process of vanishing right now, marking their final exit with small-scale explosions and releasing a puff of gamma rays. So far no one has detected any such explosions; but no one has had much time to look. No doubt they will soon, using orbiting observatories such as the Hubble Space Telescope.

RECYCLING THE STARS

Supernova explosions, which mark the end of a star's active life, also mark a beginning, for it is only by such cataclysms that the heavier elements are scattered through interstellar space, becoming the constituents of dust clouds.

By human standards, supernovae are rare things. The last one in the Milky Way seen from Earth was in 1604, and there have only been six in recorded history. By observing supernovae in other galaxies – some 300 have been seen to date – astronomers estimate our own Galaxy produces, on average, two supernovae a century (although the chances are only 50-50 that it will be visible through the obscuring clouds of dust and stars).

But in terms of the life of the Galaxy, this is not uncommon. The Milky Way experiences 20 000 supernovae in 1 million years, 20 million in a billion years. That means that the gas in any region of the Galaxy will be enriched by a supernova explosion every few million years. And from such dust clouds other stars form, as our Sun did, along with its rich and varied family of planets. The materials that form us were forged by stars, blasted into space by supernovae, and re-gathered by the newly forming Sun. Our Sun, the planets and we ourselves are literally star-stuff.

THE FAMILY OF THE SUN

2

ECLIPSED *As the Moon passes in front of the Sun, it reveals the Sun's corona.*

DURING ITS FORMATION, THE SUN, SIMILAR IN SO MANY WAYS TO OTHER STARS OF ITS SIZE, PRODUCED FROM ITS DEBRIS OF DUST AND GAS A UNIQUE FAMILY OF PLANETS, SOME OF WHICH HAVE THEIR OWN SATELLITES. EACH PLANET AND SATELLITE EVOLVED ITS OWN CHARACTER DEPENDING ON ITS SIZE, CONSTITUENTS AND DISTANCE FROM THE SUN. EACH, MOREOVER, WAS — AND IS — IN A DYNAMIC AND UNPREDICTABLE RELATIONSHIP WITH A LEFTOVER RUBBLE OF ROCKS CIRCLING IN THE SOLAR SYSTEM. ALTHOUGH THE FAMILY MEMBERS WERE FORMED FROM THE SAME RAW MATERIALS, THE MANY DIFFERENCES BETWEEN THEM THROW LIGHT ON THE WAY THE SOLAR SYSTEM EVOLVED, AND ON EARTH'S OWN PAST.

OLD SCARS *Mercury's surface is pitted by ancient bombardments.*

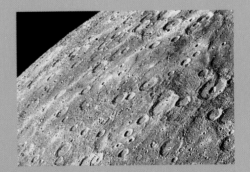

THE BIRTH OF THE SOLAR SYSTEM

Within a primordial cloud of gas and dust, a core gradually formed under the force of gravity, collapsing in on itself and heating up until it ignited in a thermonuclear reaction. The Sun was born, with a group of planets in orbit around it.

Early on the morning of June 30, 1908, a Russian peasant was taking a breakfast break, sitting beside his plough on the banks of the Angara River in central Siberia, when a noise caught his attention. 'I heard sudden bangs, as if from gunfire,' he recalled later. 'My horse fell to its knees. From the north side above the forest a flame shot up. Then I saw that the fir forest had been bent over by the wind, and I thought of a hurricane. I seized hold of my plough with both hands so that it would not be carried away. The wind was so strong it carried soil from the surface of the ground, and then the hurricane drove a wall of water up the Angara.'

When scientists arrived to investigate, they went to the site of the explosion – about 125 miles (200 km) north of the Angara, in the valley of the Tunguska River – and found devastation. There was no crater, but in a circle some 20 miles (30 km) across, trees lay felled, all splayed outwards from the centre of the blast.

For years, the impact remained a mystery. One possibility was that the object was a meteorite, a lump of rock like those that occasionally reach the Earth's surface and leave impact craters. But why would it have exploded high in the atmosphere? The most likely suggestion is that the object was a comet, weighing about 15 million tons and travelling at 12 miles (20 km) a second on a very flat trajectory – a few miles in another direction and it would have missed the Earth completely. As it was, it exploded 6 miles (10 km) up from the Earth's surface in the heat generated by its passage through the air, and released a burst of energy and a dose of radiation 1000 times greater than the Hiroshima bomb.

If this is the explanation of the Tunguska Event, as it is known, then the comet was a survivor of the Solar System's earliest days, a ball of dust and ice gathered together even before the Sun came to life or the Earth formed, circling for aeons in the depths of space until gravitational interaction

FLEETING VISITOR *Comet Hyakutake streaked across the sky in 1996, one of millions of 'dirty snowballs' that patrol the outer Solar System.*

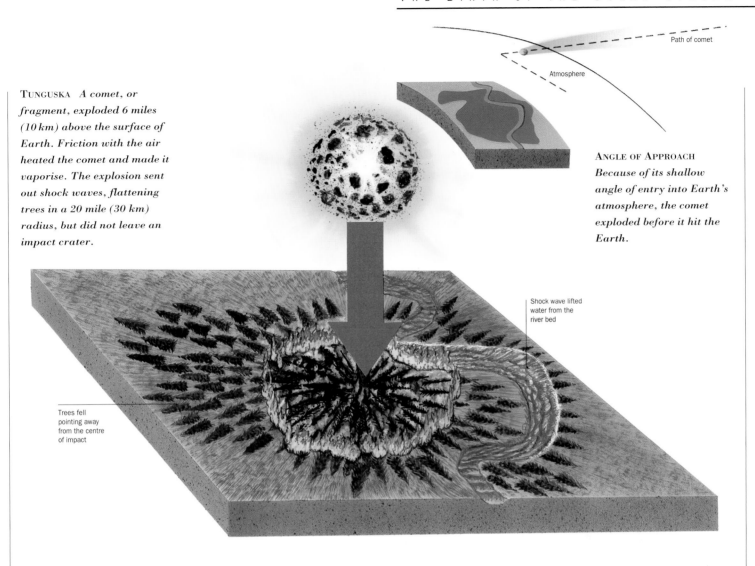

Path of comet

Atmosphere

TUNGUSKA *A comet, or fragment, exploded 6 miles (10 km) above the surface of Earth. Friction with the air heated the comet and made it vaporise. The explosion sent out shock waves, flattening trees in a 20 mile (30 km) radius, but did not leave an impact crater.*

ANGLE OF APPROACH *Because of its shallow angle of entry into Earth's atmosphere, the comet exploded before it hit the Earth.*

Shock wave lifted water from the river bed

Trees fell pointing away from the centre of impact

with some massive neighbour cast it into an eccentric orbit round the Sun, and onto a collision course with Earth. It is a reminder that our Solar System – the Sun, the Earth, the other planets and countless numbers of smaller wandering bodies – all spring from the same origins, and affect each other today as they did when they were formed.

MYSTERIES OF THE SUN

Because it is so difficult to observe directly, no one knew anything definite about the Sun's nature until quite recently. Its path (or rather, its apparent path) was recorded minutely, but its size, its distance and its workings were all mysterious. In the 3rd century BC, the Greek mathematician Aristarchus estimated the distance to the Sun by triangulation, using the Moon, Sun and Earth as points. It was an ingenious idea, with the half-Moon, exactly divided

between shadow and light, making the point of a right angle. The Sun, he said, was 20 times as far away as the Moon. The technique is fine in theory, but the measurement demands much greater precision, and he was out by a factor of 20: the Sun is 400 times as far away as the Moon.

Chinese scientists, peering at the Sun's disc through mist, noticed that it had little marks on it. As the fount of life itself, it is hardly surprising that it was equated with a god – Re in Egypt, Apollo in Ancient Greece – and it retained more than a hint of divinity until quite recently. In the Middle Ages, it was assumed that the Sun was in some way 'perfect', being circular itself and

moving in a circle, a view that came to be accepted as religious orthodoxy.

In 1610-11, Galileo turned his newly invented telescope on the Sun and noticed, as the Chinese had before him, that it seemed to be curiously marred by temporary

EARLY RECORD *A medieval French manuscript portrays an eclipse of the Sun. The Moon's shadow is shown forming a cone.*

Coronasphere

Spicules of gas caused
by Sun's magnetic field

Photosphere

Mantle of gas
500 000 K

Currents in gas
convect heat to
the photosphere

Heat radiates
outwards
8 million K

Core
16 million K

HOT SPOTS *Sunspots (the dark patches) are thought to be cooler regions on the Sun's surface where the Sun's magnetic field suppresses the hot gases rising from below.*

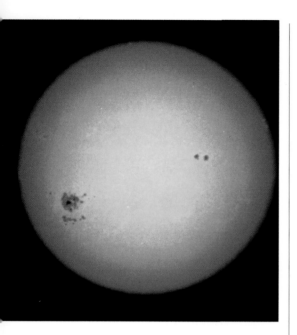

INSIDE THE SUN *Thermonuclear reactions occur at the Sun's core, where hydrogen and helium burn, and the resulting energy, in the form of heat, is transported outwards. Convection currents transport energy to the outer layers.*

spots. Moreover, it was spinning on its axis once every 25 days, not at all the perfect and unchangeable body of established teaching. But he made no guesses about how the Sun actually burned, let alone how old it was or how long it might last. Later astronomers suggested the spots might be mountains, clouds or volcanoes, as if the Sun was like a planet, but hotter.

Even at the end of the 18th century, when the scale of the Solar System and the motions of the planets were well understood, Sir William Herschel proposed that the Sun was a planet rather like the Earth, except that it had a burning outer layer of clouds. The surface, he said, was protected from the heat by an inner atmosphere of cloud that served as a firescreen. Sunspots were nothing but holes in these two cloud layers. He concluded: 'It is most probably also inhabited, like the rest of the planets, by beings whose organs are adapted to the peculiar circumstances of that vast globe.'

His words act as a reminder that the huge advances Herschel and others made in scientific observation were not matched by any awareness of how hot the Sun was or how it worked. Sunspots, those dark specks that appeared often in small groups, changed and died away, but no one understood them. The only breakthrough created another mystery. In 1851, the German astronomer Heinrich Schwabe, who counted sunspots over many years, announced that they became more and then less common, on a rough 11 year cycle, but he had no idea why.

SOLAR GENERATOR

Only in the 20th century did theory and observation combine to produce a life history of the Sun and its family of planets. That history – known as the Nebula Hypothesis – leaves a number of details unresolved, but it tells the story well enough in general terms.

Some 5 billion years ago, in an undistinguished place towards the edge of the Galaxy, a cloud of interstellar gas and dust was blasted by the expanding remnants of a nearby supernova explosion. Already, the hydrogen-and-helium cloud was rich in other elements from the burning and explosion of earlier stars over the previous 10 billion years. At some point, a random meeting of dust and gas became a seed, a core-cloud that expanded steadily as it drew in other surrounding wisps. The cloud became a ball, and as it collapsed in upon itself, as gravity drove the molecules together, it began to increase in temperature, rising from near absolute zero to –170 K after some 10 000 years, and to 1000 K after 100 000 years. Its centre contained the heat, its undefined periphery acting as a radiator. Plumes of gas carried the excess heat to what was now becoming a surface, where the gas cooled, giving off a dull glow, before gravity dragged it back towards the depths.

The gas ball's initial slight movements became a coordinated spin, which increased in speed as the ball shrank. In fact, the spin became rather complicated. Because the gas was not solid, the gas at the centre – the equator – spun faster than that nearer the poles, as it does today. The two types of movement – convection currents and spin – combined to create a dynamo that generated a strong magnetic field.

Not all the material went into what was now a protostar. The shrinking centre left behind a swirl of gas and dust, rather like the outer regions of a whirlpool. Under the combined effects of gravity and centrifugal force, these regions flattened into a disc, swirling like a liquid, with the inner and outer areas moving at different speeds, breaking up and reforming into lesser systems. This detritus would one day form nine planets, moons by the dozen and smaller bodies by the million.

As the central star gained in bulk, it trapped more and more heat in a runaway process. After about 50 million years, its core reached 8 million K, hot enough for the hydrogen to

GIANT LOOP *This huge solar flare, which occurred in 1973, spanned 300 000 miles (500 000 km). Flares, which follow lines of magnetism, release huge flows of energy that cause magnetic disturbances on Earth.*

ignite in a sustained thermonuclear reaction, creating more helium. The Sun was born, a globe 865 000 miles (1 392 000 km) across, lighting its nascent family of nine planets and their satellites.

For that family – and the Earth in particular – the important point about the Sun is the way it balances heat with stability. A central core acts as the fusion generator, burning hydrogen to helium at a temperature of about 16 million K and releasing radiation. The radiation flows outwards until it is stopped by the surrounding mantle of gas. At this point, the heat is carried away in churning convection currents of gas, forming what is known as the photosphere. The surface itself is a seething sea of hot gas, bubbling like a pot of boiling oatmeal, each bubble being over 1000 miles (1600 km) across with a life of about ten minutes. From the surface of the photosphere, jet-like spikes of gas shoot up for some 1500 miles (2400 km), driven by the Sun's magnetic field. Under the influence of magnetism, the temperature rises again sharply, to over 400 000 K, and then leaps to over 500 000 K – and sometimes up to 2 million K – to form an irregular system of radiant flares, called

the corona. Normally, the corona is invisible against the photosphere, but by coincidence during a lunar eclipse the Moon neatly covers the disc of the Sun, and the corona becomes visible for a few minutes as a glorious, dancing veil of light.

The main visible features of the Sun's surface – sunspots – are the result of magnetic storms that tear at the photosphere as the gaseous convection currents interact with the gas's surface spin. These storms also create violent discharges that bulge or shoot outwards in flares hundreds of thousands of miles long. They often curve back down again along lines of magnetic force, bridging areas of sunspot activity. The storms are unstable. The surface, spinning at different rates at different latitudes, gradually distorts the lines of magnetism until they break and reform. This happens about every 11 years, creating the sunspot cycle identified by Schwabe in 1851.

This combination of violent events – the radiation, the gaseous convection currents, massive magnetic storms – has a powerful effect way beyond the Sun's immediate neighbourhood. Blasted away from the Sun's surface by the extreme temperatures,

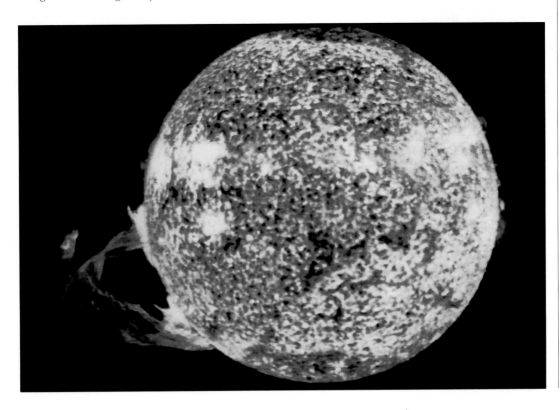

THE GREAT CRASH OF 1994

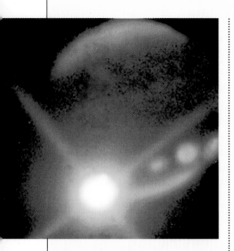

In mid-March 1993, three astronomers – David Levy, Carolyn Shoemaker and her husband Gene – at the Palomar Observatory, California, were nearing the end of a 12 year survey of asteroids and comets. The skies were cloudy, but between cloud patches they got to work. On March 25, Carolyn Shoemaker spotted a smear on the

DIRECT HIT *A major fragment of comet Shoemaker-Levy strikes Jupiter, creating a fireball some 12 500 miles (20 000 km) across.*

film near Jupiter. 'I don't know what this is,' she said, 'but it looks like a squashed comet.'

It was indeed. Research into the comet's orbit showed that Shoemaker-Levy, as it was called, was the remnants of a comet, perhaps 6 miles (10 km) across, that had been captured and torn apart into a 'string of pearls' by Jupiter's gravity in 1992. Moreover, the comet fragments would strike Jupiter in July 1994.

Astronomers everywhere, briefed via the Internet, were agog. An impact on Jupiter would be a gargantuan experiment, revealing new data about the gaseous giant. Only recently had researchers realised that comets and asteroids

played a vital role in the evolution of the Solar System, seeding planets with their chemicals, initiating rare reactions, perhaps playing a role in the beginnings of life, and certainly being a significant factor in great extinctions. Here was a once-in-a-millennium chance to observe an actual impact, with wonderful new tools – the Hubble Space Telescope and the Galileo spacecraft, then en route for Jupiter.

For over a year the tension grew. Some scientists feared a non-event, a mere fizzle, while tabloid papers revelled in predictions of a cataclysm. Doubts and hype dissolved on July 16, however, when the spectacular impact occurred.

By then, the comet had split into more than 20 fragments ranging from 330 ft (100 m) to an estimated 2$\frac{1}{2}$ miles (4 km) across, stretched over 900 000 miles (1.5 million km). The first fragment produced an

IMPACT SITE *The dark patch near the edge of Jupiter marks one of the impact sites, photographed by the Hubble Space Telescope.*

awesome explosion. It struck on the planet's limb, just out of sight from the Earth, at 37 miles (60 km) per second. A fireball burst in Jupiter's atmosphere and exploded upwards for 1865 miles (3000 km) after only five minutes. Over the next 20 minutes, the explosion ballooned out into a plume 6000 miles (10 000 km) across, producing an immense dark scar that swung into full view as Jupiter spun.

For a week, until July 22, the rest of the fragments rained down onto Jupiter, some dissipating with no explosion, some blasting a line of immense, dark spots into the atmosphere. These clouds, the dusty remains of the fragments, were slowly smeared out. After eight months, scarcely a trace remained of the grandest event in the Solar System ever witnessed by humans, but the evidence gathered will take many years to assimilate.

IMPACT RIPPLES *A $\frac{1}{2}$ mile (1 km) fragment of Shoemaker-Levy affects the whole Jovian atmosphere. The comet broke into more than 20 fragments.*

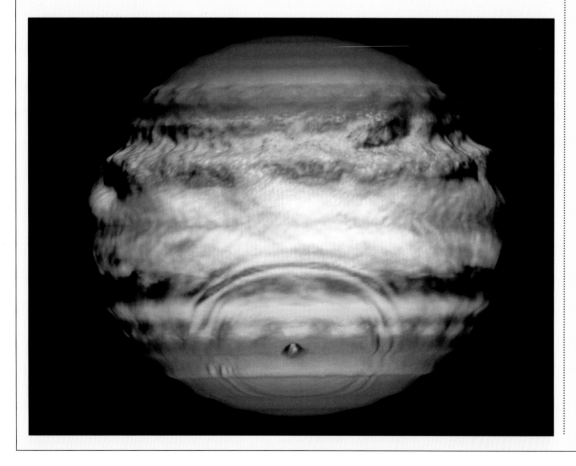

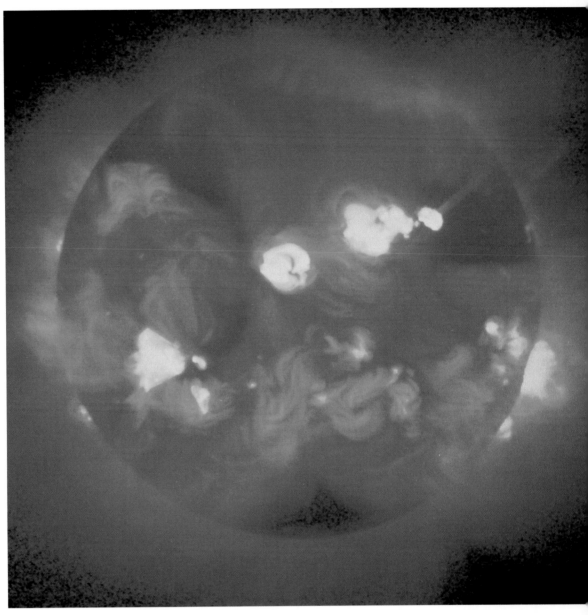

SOLAR CORONA *An X-ray image of the Sun shows the intense activity of the corona, the outer layer of the Sun's atmosphere. The bright areas are sunspots on the surface.*

gas flows outwards forming the so-called solar wind, a tenuous plasma that carries with it a faint echo of the Sun's magnetic field and thus creates a diffuse, magnetised sea through which the planets swing. The massive magnetic storms on the Sun make waves of their own in this sea. Five days after a large-scale flare on the Sun, the radiant energy hits the Earth, directed north and south by the Earth's own magnetic field, causing the great flickering sheets of northern and southern lights, often disrupting communications and even causing changes in the weather.

The Sun is not an everlasting source of power, being now about half way through its life. It has been burning much as we now see it for almost 5 billion years, and it will go on burning for another 5 billion years before its hydrogen fuel begins to run out.

THE SOLAR FAMILY

This general view explains many things about the Solar System. For instance, several characteristics seem to derive from the planets' origins in a disc of gas and dust spinning round the Sun. The planets all orbit the Sun in the same direction, which is also the direction of the Sun's spin. Most of them orbit on roughly the same plane but the orbits of Mercury and Pluto are tilted slightly – by angles of 7° and 17° respectively. With the exception of Mercury and Pluto, all the planets have almost circular orbits. Many comets – assumed to be leftover fragments from the very early days of the Solar System, before the disc had formed properly – come

from every part of the sky, not simply the plane of the disc.

These common traits, though, conceal several mysteries crucial to an understanding of the Solar System's origins. The inner planets – Mercury, Venus, Earth and Mars – are small and solid. Of the five outer planets, four (Jupiter, Saturn, Uranus and Neptune) are gaseous giants. Why the differences? Why should the last, Pluto, be a twin of Mercury, and why should it be in a 'wild' orbit? And why should three planets (Venus, Uranus and Pluto) rotate in the opposite direction to their movement around the Sun?

The most crucial problem concerns the rotational movement, or angular momentum, of the Solar System. To understand the problem, recall the image of the spinning skater speeding up by drawing in her arms. In effect, the Sun did this as it contracted under the force of gravity. Unfortunately for theory, it did not obey the laws of physics. Too much angular momentum was retained by the planets (99 per cent), and far too little (the remaining 1 per cent) by the Sun. It 'should' be spinning 400 times faster.

The most likely explanation is that the Sun's rotation was slowed by its magnetic

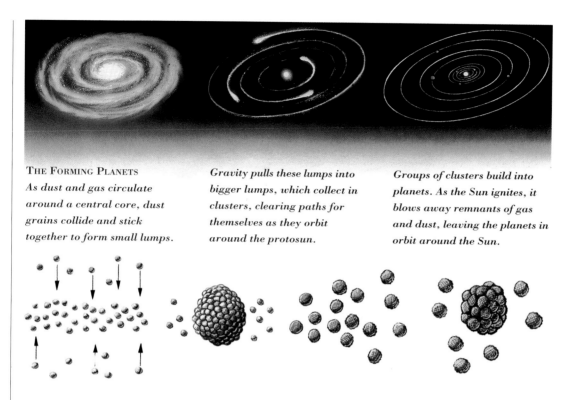

THE FORMING PLANETS

As dust and gas circulate around a central core, dust grains collide and stick together to form small lumps.

Gravity pulls these lumps into bigger lumps, which collect in clusters, clearing paths for themselves as they orbit around the protosun.

Groups of clusters build into planets. As the Sun ignites, it blows away remnants of gas and dust, leaving the planets in orbit around the Sun.

similar to the one that produced the Sun. Before the Sun had ignited, in many thousands or millions of places in the disc local disturbances forced a few grains to collide. Since any grain even a few yards farther from the Sun orbited at a slightly different speed, the inner grains overtook the outer ones. Interacting, drawing closer, colliding, they formed small flakes, then ever larger pieces, from bodies the size of gravel to those the size of stones, rocks and mountains, sweeping a clear zone for themselves through the disc. It was a random business, because as the objects grew in size and collided, their impact must sometimes have been so great that one or other, or both, would have broken up. Finally – after perhaps 100 million years of accretion, collision, fracture and re-formation – the accumulated material had been assembled into very rough planets.

field interacting with the solar wind. Rather like a person trying to spin in water with arms outstretched, the Sun was subject to drag from its own products. The drag took effect long before the planets emerged, because when they formed they reflected the angular momentum the Sun had once possessed.

Is it possible to explain how the planets, and their regularities and discrepancies, formed? The short answer is: not entirely, at least not yet. But astronomers believe they have the general idea.

First, how did a disc of gas and dust produce solid bodies? Probably by a process

WANING INFLUENCE *The Sun exerts less influence over the outer planets than the inner ones.*

Heat from the Sun decreases with distance

Too hot to sustain life

Too cold to sustain life

Average surface temperatures:

470°C (878°F)	150°C (302°F)	10°C (50°F)	−53°C (−63.4°F)
Mercury	Venus	Earth	Mars

Jupiter

Saturn

−214°C (−382.2°F)

−218°C (−360.4°F)

Uranus

Neptune

Pluto

—— TERRESTRIAL PLANETS ——

Material with a low boiling point has been vaporised by the Sun's heat and pulled away by the Sun's gravity

—— GASEOUS PLANETS ——

As heat and gravitational pull from Sun are low, these planets have kept their thick, icy atmospheres

Strength of gravitational pull increases towards Sun

The process varied with each planet's distance from the forming Sun. This was a process in which temperature imposed character, and each planetary core became its own test tube with its own chemical reactions. Near the hot centre of the disc, where the temperature was already perhaps 2000 K, nothing could stick together. Between about 50 million and 200 million miles (80-300 million km) out, where the temperature fell from 1400 K to 400 K, gases would still be kept in unrestrained motion by the heat, but solid grains of dust could form, with their high content of elements. Four lumps rich in nickel and iron formed – the cores of the four inner planets, though with each jump from the Sun, the core tended to acquire ever lighter materials. In the colder outer regions of the disc, elemental grains mixed with volatile gases to form the cores of what would become the giant gaseous planets.

Then, at last, the Sun's nuclear fires were ignited, blasting the still-forming system of planets with radiant energy. Most of the remaining gases were blown away from the inner planets by the solar wind. Under this assault, and now under the influence of their own trapped radioactive heat, the cores of the inner planets melted and condensed. Less dense materials – such as silicates – rose to the surface to form a mantle and an outer skin, or crust. Farther out, the gaseous giants, which would otherwise have been vulnerable to such an outpouring of energy, were protected by their distance and were free to develop in their own way.

INDEPENDENT BODIES

Once the planets and their satellites were formed, a vast mass of unused material – dust particles, rocks, drifting gas – was still in the Solar System. The heavier of these, some in regular orbits, some swinging wildly between the planets, remained as asteroids. Lightweight gaseous balls were blasted into the outer reaches of the Solar System, way beyond Pluto, possibly forming a vague, very distant and diffuse mass – known as the Oort Cloud after the Dutch astronomer Hendrik Oort, who proposed its existence. Even with today's techniques the Oort Cloud cannot be seen, but many astronomers believe that it does exist and that it is the most likely source of comets.

The detritus of rock, dust and gas was of particular significance, for the process of formation did not stop. The Sun and planets continued to sweep up the leftover pieces. Some of the rubbish fell into the Sun. For the planets, the next few hundred million years was filled with drama and near catastrophe, with each planet subject to an irregular bombardment, each new impact blasting a new crater, sometimes splashing molten rock far into space.

Only after about 600 million years did the Solar System settle into something like stability. Mercury and the Moon, now dead worlds in that their internal fires have died and changes to their geological structures and surface have ceased, still show evidence of the bombardment that ended some 4 billion years ago. The evidence is also there, perhaps, in other irregularities, like the eccentric orbit of Pluto and the 'reversed' spin of three of the planets.

Over the course of 4 billion years, most of the dust and the larger bodies have been blotted up. But there is still a good deal of both left, falling as a sort of 'hard rain' and providing evidence about the earliest days of the Solar System. Most of this rain consists of droplets – dust particles – that arrive in Earth's atmosphere in stupendous numbers, perhaps as many as 100 million each day. Most are smaller than pinheads and burn up unseen, but some slightly larger ones penetrate farther.

Whenever a large speck of dust falls into the upper atmosphere, moving at several thousand miles an hour, it heats up and

ASTEROID BELT *Many asteroids circulate in a belt between Mars and Jupiter. Pluto's 'wild' orbit sometimes brings it within Neptune's orbit (bottom left).*

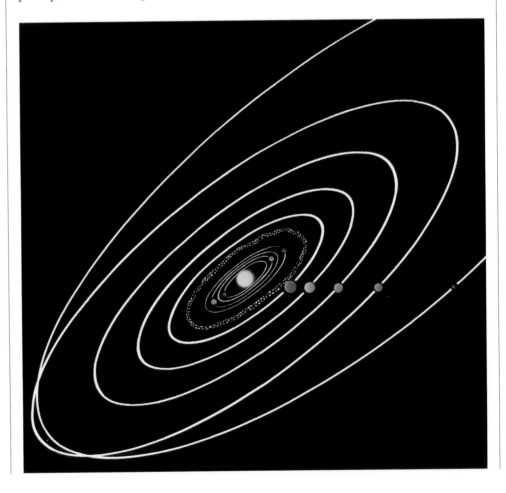

vaporises in a flash of light – a 'shooting star' or meteor. Anyone watching on a clear night can expect to see about half a dozen an hour. The term 'meteor' derives from the Greek word covering all atmospheric phenomena – it was not until the 19th century that 'meteorology' became restricted to the study of weather. At times the circling specks arrive in swarms, creating 'meteor showers', like the Perseid shower that streams in from the direction of the constellation Perseus in August. Usually showers are annual – there are over a dozen every year – but occasionally they are one-offs. The year AD 902 was nicknamed the Year of the Stars because meteors streaked across the sky as thickly as snowflakes.

Most of this 'hard rain' comes from comets, lumps of ice and dust. At first sight, there is nothing very hard about a comet. The head consists mainly of loosely bound grains, forming what astronomers often refer to as a 'dirty snowball'. Most of the time they circle in the icy interstellar wastes, so far from the Sun that the Sun seen from a comet would appear not much brighter than any other star. Only when they fall towards the Sun do they become true

REGULAR VISITOR *Edmund Halley used Newton's newly formulated laws of gravity to predict the return of the comet named after him.*

comets, drawn by the Sun's gravity, accelerating in, then swinging round in a tight turn to vanish outwards again. Some circle close in – Encke's comet returns every 3.3 years. Some interact with a planet or two and are slung for ever out of the Solar System. Some, ranging out halfway to the nearest star, may take as long as 10 million years to make a single circuit of the Sun. As yet, no one can estimate how many comets may lurk out there in the Oort Cloud – estimates range up to 100 billion – because only a tiny fraction ever fall inwards, nudged towards the light by the faint gravitational effects of a star, gas cloud or passing asteroid.

On their journey, comets take the shape that has been familiar to humans from the earliest historical times: a tiny head and the streaming tail from which they derive their name – *aster kometes*, 'long-haired star', in Greek. Until the late 17th century, they were wonders, appearing from

RECORDED FOR POSTERITY
The appearance of a comet – now identified as Halley's – in 1066 was recorded in the Bayeux tapestry.

nowhere, explicable only as signs and portents. When one appeared in 1066, just before William the Conqueror's invasion of England, he must have wondered what it presaged. Soon he knew the answer: victory. After the Battle of Hastings, the omen was remembered and embroidered into the Bayeux tapestry.

Six hundred years later, in 1682, another comet appeared. The astronomer Edmund Halley spotted an extraordinary opportunity. He had recently published, at his own expense, the laws of gravity worked out by his friend Sir Isaac Newton. If those laws were right, then the comet would be subject to them. He consulted past records, found a pattern, and announced that this comet was in orbit round the Sun, taking about 76 years to return. Halley's research was the first prediction based on Newton's laws – and when the comet returned in 1758, after Halley's death, the laws received a brilliant vindication. It was then simple to calculate that Halley's comet and the Bayeux tapestry comet were one and the same.

From then on, comets were stripped of their ominous nature and became objects

THE GREAT BOMBARDMENT

Impact craters, where one object left its exact trace on another, are present on the inner planets and on several distant satellites, revealing the past life of the Solar System when rocks were formed and re-formed by collisions. Some have been eroded by later bombardments and by the geological activity – earthquakes, volcanoes, melting, freezing and chemical change – that marks each body as an individual.

Impact craters on Mercury, a small, long-dead planet, are similar to those on the Moon. All are ancient remnants of collisions that took place over 3.5 billion years ago. One huge feature, Caloris Basin, looks like a lunar *mare*, or sea. Some 800 miles (1300 km) across, it is filled with lava melted by the impact, which flooded craters made by earlier impacts. Mercury has retained a regular scattering of later, smaller craters, but fewer of them.

Craters on Mars are scattered unevenly, with more in the southern hemisphere, revealing that the north was subjected to a greater number of volcanoes and lava outflows, which filled in more ancient craters. The southern hemisphere resembles the Moon and Mercury, with a few huge craters and many smaller ones. A number of these are filled with wind-blown dust that has eroded the crater walls and been piled into dunes. In addition, Martian craters lack rims of ejected material. Instead, they have bulging rims that were probably created by meltwater and mud released from beneath the surface on impact and almost instantly frozen. Even Mars's two little moons, Phobos and Deimos, are pockmarked with small craters. A dense atmosphere, vicious winds and massive volcanic activity have re-formed the surface of Venus, but several massive craters have survived. One, called Mead (after the anthropologist Margaret Mead), is 170 miles (275 km) across, and forms a double ring filled with lava.

SATURNINE SATELLITE *One of Saturn's moons, Enceladus, has an unstable icy surface and a smattering of craters, proof of recent bombardment.*

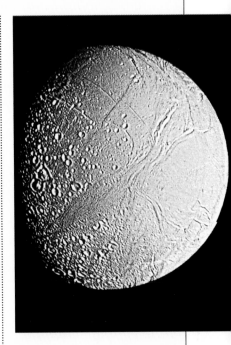

Ganymede, Jupiter's largest moon, has an ancient, stable and well-cratered surface, with craters ranging up to 95 miles (150 km) across. The craters are shallow compared to those on the Moon and Mercury, however, indicating an icy rather than a rocky surface, and one that flowed back into the craters after impact, partially refilling them.

Callisto, the farthest out of Jupiter's inner family of so-called Galilean moons, is riddled with craters, some with bright rays of shattered ice, some filled with ice. One enormous impact crater, Valhalla, has concentric rings of distorted rock 1600 miles (2600 km) across, and a bright, shattered bull's-eye some 370 miles (600 km) across. The feature was created by a stupendous collision that sent melted rock and ice flooding outwards, only to be quick-frozen in the –165°C (–265°F) surface temperature.

All 18 of the known moons of Saturn are cratered except the largest, Titan, which has a dense atmosphere. The strangest is Enceladus, with a radius of only 155 miles (250 km). Its crust has only recent, small craters up to 20 miles (35 km) across, larger and older ones having been ironed out by a crust deformed by internal heat.

Of the 15 known moons of Uranus, several have a mixture of strange and cratered features. Miranda seems to be made of rough, dirty ice, with very few craters. Titania's crust became static aeons ago, and has an even scattering of ancient craters. Ariel's surface is newer, with lighter craters. On Umbriel, the craters are so densely packed that they overlap.

SCARS OF BATTLE *The first close-up view of Mercury, taken by the Mariner probe in 1974, reveals a surface as battered as that of the Moon.*

of research. Hundreds are now known, Halley's best of all. In 1986, when Halley's comet returned again, the European Space Agency's *Giotto* spacecraft swept in to within 375 miles (600 km) of the core, and found it to be a dark, peanut-shaped object measuring about 5 × 10 miles (8 × 16 km), with a mountain and several craters. The craft also discovered that Halley's tail included some surprisingly complex chemicals, such as sulphur and carbon compounds, which are basic constituents of life – proof that Earth shares a common origin with comets.

It seems strange that comets should have a tail, for there is no air to create a 'wake'. Yet there is a sort of atmosphere – the solar wind. Radiant energy from the Sun strips away a tenuous spray of dust from the head and scatters it downwind. For this reason, as the comet swings round the Sun and heads off again into the abyss beyond Pluto, its 'tail' precedes it, pushed by the solar wind. Comets are a triple source of 'hard rain'. Their tenuous tails form slightly denser clouds of interstellar dust, the stuff of meteor showers. The Perseid shower is linked to the comet known as Swift-Tuttle, discovered in the 19th century and rediscovered in 1992. Swift-Tuttle comes by once every 130 years, but its debris remains as a sort of fossilised wake, and the Earth sweeps through it every year.

Comets can be torn apart by the influence of a planet's gravity and swept to dramatic ends. In 1826, a new comet was found with a period of 6.75 years. By 1845, three passes later, it had broken in two. By 1872 nothing was left of it, but some of it fell to Earth in a spectacular meteor shower. Undoubtedly the most impressive break-up occurred in the early 1990s, when a new comet was found, and named Shoemaker-Levy after the two astronomers who identified it. The comet, caught by Jupiter's gravitational field, broke up into an extended line of subcomets that finally

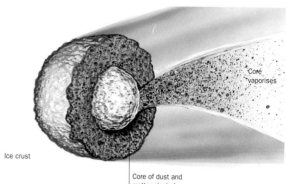

CONSTRUCTION AND ORBIT OF A COMET *Heat from the Sun melts the comet's crust, and a cloud of dust and grit is ejected. Eventually, the cloud breaks down forming a stream of dust behind the comet and a gas tail.*

Core vaporises

Ice crust

Core of dust and matter ejected

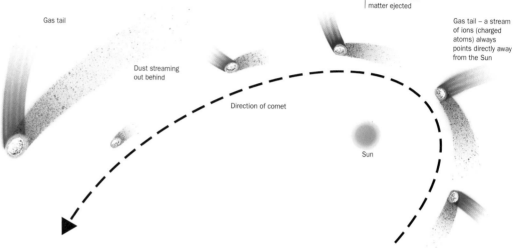

Gas tail

Gas tail – a stream of ions (charged atoms) always points directly away from the Sun

Dust streaming out behind

Direction of comet

Sun

fell into Jupiter in July 1994 with a series of massive explosions.

Comets also create hard rain when their volatile gases are blown away by multiple passes round the Sun, leaving a chunk of rock, like millions of others in orbit between the planets.

METEORITES

Pieces of rock orbiting in space, most of which are too dense to have derived from comets, are most commonly the size of a pebble or small stone, but they can be larger, and occasionally are very large indeed. Whatever their size, those that fall to Earth are known as

meteorites. Almost since the dawn of history, people have recognised their celestial origin and often venerated them as divine. In the Bible, the Ephesians worshipped the goddess Diana and 'the thing that fell from the sky', while the Muslim holy of holies, the Black Stone in the Grand Mosque, Mecca, is almost certainly a meteorite, blackened by its passage through the air. Since meteorites were traditionally regarded as magical, scientists dismissed them as objects of study until 1794, when a German physicist, Ernst Chladni, overcame the ridicule of his peers and argued convincingly that meteorites were pieces of cosmic matter.

Every year, hundreds of little ones plummet to Earth, with a flash and sometimes with a series of explosions. Though dull-looking to untrained eyes, they have great importance for astronomers because they are the only pieces of the Solar System

CHILEAN METEORITE *This stony-iron meteorite, pitted from its burning descent through the atmosphere, fell in the Atacama Desert in 1823.*

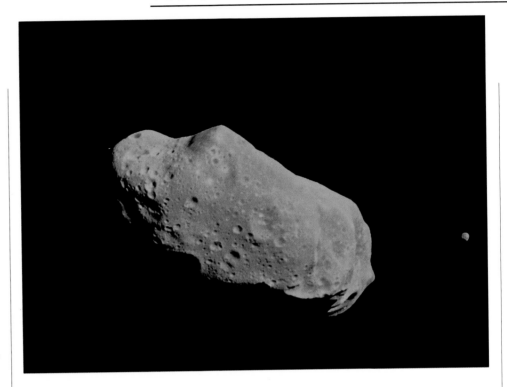

that can be studied directly, other than Moon rocks brought back by US astronauts on the Apollo missions. Most meteorites fall into the sea, but a few dozen are retrieved every year, and occasionally people are subjected to near-misses. So far, no one has been reported killed by a meteorite, which is an indication of their relative rarity.

Meteorites can be puzzling things. They may be predominantly of stone or iron, some have a high carbon content, and even water, depending on how far from the Sun they originated. The structure of some that are rich in nickel and iron – the same material as the cores of the inner planets – suggests that they cooled very slowly, just 1°C (1 K) every million years. So they could not always have been their present size. In fact, they must have been contained within a blanket of rock at least 60 miles (100 km) across. These mini-planets were large enough to generate their own internal heat and develop a core and a crust. Over millennia, the crust was smashed by collisions with other bodies, leaving an iron-rich core.

Those that make it to the Earth's surface can have a dramatic effect. Meteor Crater in Arizona, 600 ft (180 m) deep and ³/₄ mile (1.2 km) across, was blasted out by a chunk of rock some 20 000 years ago. Wolfe Creek is the largest and most spectacular of five clearly recognisable craters made by meteorites in Australia, and was

METEOR FROM MARS *This sliver of meteor is from a rock that was possibly blasted from Mars by a massive impact and then fell to Earth.*

created by the impact of a meteorite weighing more than 50 000 tons. A 2 ton meteorite fell near Pueblito de Allende, Mexico, in 1969. In fact, it was from this meteorite that scientists were able to derive a good deal of information about the early Solar System, for it contains a peculiar form of magnesium that could only have been produced within a supernova. The magnesium must therefore have been blasted across interstellar space before being incorporated into the Allende meteorite, so here, in a sort of interplanetary fossil, was evidence of the event that could perhaps have triggered the formation of our Solar System.

ASTEROIDS

Large meteorites begin to rival in size the final category of free-ranging objects, the asteroids, which vary from huge, irregular-shaped rocks to roughly spherical objects the size of small moons. The largest, Ceres, discovered on the first day of the 19th century, is about 600 miles (1000 km) across. There are uncounted thousands of others. Many lie in the space between Mars and Jupiter – the raw material, perhaps, of a planet that never formed. Some sit in two little bunches called the Trojans in Jupiter's

FLOATING BOULDER *The 35 mile (56 km) long asteroid Ida and its moon were photographed by the probe Galileo in 1993.*

orbit. Others swing in eccentric orbits, occasionally approaching dangerously close to Earth. Like the planets, they almost all fall within the plane of the planetary disc, which suggests that they were formed by the same process, though one – Icarus – cuts in from below the disc, as if it were once ranging freely between the stars and was captured by the Sun.

The distinction between asteroids and meteorites is only one of scale. They are made of the same stuff, and originated in the same processes; asteroids being larger, are rarer – and correspondingly more dangerous should they ever approach a planet. Around 100 bodies more than ¹/₂ mile (1 km) across are circling the Sun in orbits that cross the Earth's, but there are undoubtedly hundreds of others unrecorded.

Randomly, they strike. Once or twice a century, something as big as the Tunguska body hits the Earth. Every 500 000 years or so, the Earth, like any other planet, can expect something much worse – all reminders that the Earth is still evolving in the same dynamic environment that shaped its evolution, and that of all its siblings.

THE EARTH'S SIBLINGS

The family of planets that make up our Solar System were all formed from the same basic elements, and have many features in common. Yet they have evolved into very different places, from lifeless worlds of stark beauty to restless gaseous giants.

As the solar family evolved from its promising environment – Sun, radiant energy, dust, gas and assorted scatterings of rock and ice – it emerged as two subgroups: small, dense planets near the Sun, and large, soft, gassy ones farther out.

Recently, that simple view has been complicated by the discovery of just how individual the planets are, for example in their numbers of satellites: the count for the whole Solar System now stands at 61. But every new discovery about Earth's siblings adds to our knowledge of the framework within which our planet developed, and what happened to prevent the evolution of any other Earth-like planets. At times luck, in the form of random collisions with stray asteroids, played a part, but there is an underlying logic: only the Earth is just right. None of the others are able to support life – too close to, or too far from the Sun; too big or too small; too hot or too cold.

THE INNER PLANETS

Take Mercury, the planet nearest to the Sun. It has many peculiarities; it is bright, fast-moving, and enigmatic because it is almost impossible to examine against the Sun's blinding light. Then quite suddenly, in 1974, the veil of mystery was raised when Mariner 10 flew past and made a detailed photographic record. But the solutions to one set of mysteries raised others. There is a rather odd relationship between Mercury's orbit-time (almost 88 days) and its rotation time (about 59 days): it rotates only three times while going round the Sun twice. Its day lasts two Earth-years. It also has a rather eccentric elliptical orbit, 43.5 million miles (70 million km) long by 28.5 million miles (46 million km) wide.

Perhaps these oddities were all caused by the same ancient event – an encounter with a particularly large asteroid, which left a scar, Caloris Basin, some 800 miles (1300 km) wide. The impact set up such a shock wave that it jumbled the surface immediately opposite the impact point. It might also have been enough to bump Mercury into its eccentric orbit and melt so much rock that it acquired tides. Perhaps the friction of these rocky tides, slopping back and forth as Mercury cooled down again, slowed the planet's spin until it fell into its unique 3:2 harmonic relationship with its orbit time.

In its history and structure, Mercury has traits of both a planet and a moon. It seems to have formed as the Earth did, with a hot core, for it has a magnetic field, probably produced by the dynamo at its centre –

NEIGHBOUR TO THE SUN
The heavily cratered surface of Mercury was photographed from the space probe Mariner 10 in 1974.

a liquid iron core spinning as the planet spins. But it is Moon-sized, being only about 40 per cent larger – 3000 miles (4900 km) across – and equally pockmarked with ancient craters. Like the Moon, it has no atmosphere to speak of. Nor does it show signs of volcanic activity.

A closer look, however, reveals differences from the Moon. Though it has several huge craters over 125 miles (200 km) across, Mercury has fewer craters in the 12-30 mile (20-50 km) range than the Moon. It has no mountains, more smooth plains, and masses of long, scalloped escarpments. Apparently, when the Solar System was forming, meteorites and asteroids were drawn away from Mercury by the overwhelming force of the Sun's gravity. And, unlike the Moon, Mercury seems to have shrunk while cooling, acquiring its wrinkle-like escarpments. Why the similarities, and why the differences?

Some are due, perhaps, to the massive impact that made the Caloris Basin, which was probably enough to amputate a good deal of Mercury's mantle, leaving it permanently crippled. But Mercury's character derives mainly from its small size. It does not have the gravitational force to hold down an atmosphere, and it does not have the size to retain its internal heat. This means that it lacks any mechanism that would create volcanoes, earthquakes or mountains, or create new chemical combinations, let alone release them at the surface. Soon after its formation some 4 billion years ago the fire within Mercury died, and the planet became the passive victim of a declining bombardment of rocky debris.

VENUS: ALMOST ANOTHER EARTH

Moving out from the Sun, the next planet is Venus. It might have been Earth's twin, for it is almost the same size, and was probably formed in the same way. Along with Earth and Mars, Venus is the innermost of the three so-called terrestrial planets. Its origins and destiny have special relevance for us.

Veiled in clouds, bright as a beacon, Venus had a soft, even romantic image, fitting for the classical goddess of love. Since her face was hidden, scientists and authors

were free to dream, fantasising that perhaps the clouds concealed vast tropical rain forests, swamps or oceans. In the 1950s, the millions of British children who were avid readers of the *Eagle* comic were enthralled by the adventures of Dan Dare, 'pilot of the future', and his evil Venusian rival, the Mekon. In that steamy world, almost anything might be possible.

Radar studies in the 1960s, and a score of later spacecraft investigations, shattered the dreams, but revealed a world far more extraordinary than any science-fiction fantasy. The first surprise was that

BRIGHTEST PLANET *Venus's permanent cloud cover reflects 79 per cent of sunlight, making it the brightest object in the sky after the Sun and Moon.*

Venus rotates extremely slowly, taking 117 Earth days to go round once – too slow to produce the dynamo effect at the core that generates a magnetic field. More than that, it rotates *backwards*. On Venus, the Sun rises in the west and sets in the east.

In 1975, in one of the greatest technological successes of the space race, the Soviet

Incoming radiation (energy from the Sun) passes through the atmosphere

Reflected heat energy is trapped by carbon dioxide

Energy heats the ground

GREENHOUSE EFFECT *Venus's rocky surface absorbs sunlight and radiates heat, which is trapped by carbon dioxide in the atmosphere.*

Union landed two Venera probes, which sent back the first pictures from another planet. The probes were equipped with searchlights in case the cloud cover veiled the surface in stygian darkness. In fact, they revealed rocks and sand under a sky 'as bright as a cloudy day in Moscow'. They also revealed that Venus is anything but a romantic place. Rather, it resembles a medieval view of Hell. The ground is barren, the atmosphere is carbon dioxide – like car exhaust emissions – at 90 times the pressure of Earth's, and its shield of clouds is made of sulphuric acid droplets. On the ground, the temperature is 600°C (1112°F), enough to make metal glow red hot. Though two later landers did better, those first two were roasted to ashes in an hour.

In August 1990, a US Magellan spacecraft in orbit began sending back a detailed radar map of Venus's surface. It is scattered with many volcanoes, all dead. There is no sign of moving, Earth-like continental plates, but some hints of local movement – rocks that have been folded or broken apart. There are also numerous impact craters, many of them filled with lava. It seems the whole surface was reformed violently no more than a few hundred million years ago, making the surface only about twice as old as the Earth's. The crust is remarkably uniform, like dough, not stiff enough to slide around, but wrinkling and puckering under the influence of volcanoes.

With this information, it is possible to sketch Venus's history. Like Earth, it grew by accretion, developed a core and a mantle, and endured aeons of concentrated bombardment by asteroids. Presumably, in the early days of its formation, a particularly

UNDER A BRIGHT SKY *This computer-generated view shows Venus's grandest mountain range, the 7 mile (11 km) high Maxwell Montes.*

awesome collision with an asteroid reversed the normal direction of spin.

The key difference with Earth, though, was the lack of liquid water. Originally, Venus had enough hydrogen and oxygen to make water, but instead of combining to form a liquid, the gases remained separate, along with carbon dioxide. Possibly other elements – iron and sulphur – absorbed the oxygen, while the lighter hydrogen escaped into space, leaving carbon dioxide. It was the free carbon dioxide that came to govern Venus, for it is a 'greenhouse gas' that traps infrared radiation, or heat. As carbon dioxide was blown out by volcanoes, it triggered a runaway 'greenhouse effect'. Venus, a twin of the early Earth, became an exhausted planet, while the Earth evolved along different lines. Ecologists like to draw conclusions about the possible fate of the Earth if carbon dioxide from pollution is allowed to build up indefinitely in our atmosphere.

MARS: A LIVING WORLD THAT DIED

Of all Earth's neighbours, Mars is the one that offered the strongest evidence of being Earth-like. Its surface could be seen directly from Earth. Its reddish tinge showed variations, and its white, icy poles shrank and

grew, which suggested the existence of weather and seasons. It was a close neighbour, only half the size of the Earth, but with a spin about the same as ours. There was enough here to suggest intriguing similarities and inspire one of the oddest controversies in the history of astronomy.

The controversy concerned the 'canals' of Mars. The idea was first popularised in the 1880s by the Italian astronomer Giovanni Schiaparelli, after he mapped what seemed to be strange lines, some of them thousands of miles long, crisscrossing the surface of the planet. He called them *canali* and suggested they were geological features formed by water flowing from the icy poles.

The idea inspired a wealthy American, Percival Lowell, to establish his own observatory in Flagstaff, Arizona, in 1894 and dedicate most of his life to the subject. As if obsessed by an anglicised version of the Italian word, he became convinced that the

MARTIAN MYSTERY
The surface of Mars is reddened by its own dust. The markings once mistaken for canals are clearly visible.

canali were actual canals, artificial waterways constructed by Martians. He also guessed that Mars was a desert planet, and suggested that the Martians were struggling to irrigate their dying world. He explained that the

lines were made by the canals and the vegetation that grew along them, showing up rather as the Nile appears from space. He concluded: 'A mind of no mean order would seem to have presided over the system we see.'

In the 1960s, the coming of the Space Age added a couple of twists to the saga. Lowell's ideas were proved fanciful, based on optical illusions and wishful thinking. Neither Schiaparelli nor he could possibly have seen such fine detail with an Earth-based telescope. Yet, strangely, there was something behind the idea. When the first of four Mariner space probes took close-up shots, they revealed that the planet did indeed appear to have had running water, once upon a time. There are 'canals' after all, invisible from Earth and dry for aeons, the existence of which throws light on the evolution of the Earth.

Much else emerged to suggest that Mars was once much more Earth-like, but not in terms of its present-day biochemistry. The first pictures of the surface, taken by the US Viking soft-lander in 1976, showed a drab, dusty desert landscape, devoid of activity. Experiments designed to search for signs of life found nothing – not surprisingly, for its atmosphere is mainly carbon dioxide, at a

pressure only five-thousandths that of Earth's. There is enough water vapour to form the polar caps, but not enough to flow. It is now known that if all Mars's water were gathered together in liquid form, it would create just one large lake. There is no rain. In the thin air, the temperature seldom rises above freezing. The surface down to 1 mile (1.6 km) deep is probably permafrost.

Yet since then, large-scale photographic coverage from orbit has revealed that Mars is still a dynamic world. It has huge volcanoes, including the largest so far known in the Solar System, Olympus Mons, 15 miles (25 km) high (three times the height of Everest) and 375 miles (600 km) across at its base, with a crater the size of London. A great rift valley, evidence of an active crust, cuts across the northern hemisphere. Gales whip up violent dust storms, forming and re-forming sand dunes. Impact craters from its early history have been eroded by sandstorms and filled in by lava flows.

The most intriguing features, though, are the ravines. There seems little doubt that they were once watercourses. Like dry riverbeds in deserts on Earth, they run downhill, they meander, they have tributaries and they created outwash plains. The watercourses – some of them as large as major rivers on Earth, hundreds of miles long – all stem from highlands that have

DUSTY LANDING *The surface of Mars, captured by a Viking lander in 1977, is an inhospitable place of volcanic boulders and drifting dust.*

been shattered by movements of the planet's crust, and from areas that seem to have collapsed, as if undermined by water flowing underground.

All of this suggests the following possible biography for Mars. It formed in the same way as Earth, with a crust, a mantle and a core. Asteroid impacts blasted the surface, churning it up to a depth of 1 mile (1.6 km) or more, releasing enough oxygen to make water, a process aided by volcanoes. The amounts of water would have been massive, literally oceanic. According to one estimate, there would have been enough water to cover the whole planet to a depth of at least 260 ft (80 m), though production would have been spread over millions of years, so it is hard to guess how much water existed at any one time. In any event, Mars could well have had a dense atmosphere and millennia of rain. Much of the water percolated underground, where it formed reservoirs.

Volcanic action continued, and it was this that released the ground water. But now Mars's size proved crucial. Its gravity could not hold its gases in place, and they escaped into space. As water was released it evaporated, except at the frozen poles. The water continued to pour out until the supply was exhausted, perhaps relatively recently – only hundreds of thousands of

continued on page 64

PLANET STRUCTURE *Terrestrial planets have large rocky cores, whereas the gaseous planets have small rock or ice cores surrounded by layers of gas.*

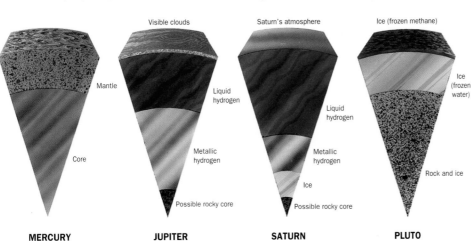

MERCURY
Mantle
Core

JUPITER
Visible clouds
Liquid hydrogen
Metallic hydrogen
Possible rocky core

SATURN
Saturn's atmosphere
Liquid hydrogen
Metallic hydrogen
Ice
Possible rocky core

PLUTO
Ice (frozen methane)
Ice (frozen water)
Rock and ice

VOLCANOES ON OTHER WORLDS

Volcanoes are a sure sign of geological life. Four bodies besides Earth – two planets and two moons – are known to be active in this way.

Mars has the largest volcanoes in the Solar System. The biggest of all is Olympus Mons, 375 miles (600 km) across at its base, and an astonishing 15 miles (25 km) high, three times the height of Everest. The lava flows from it suggest that the mountain last erupted 1 billion years ago. But Olympus and the volcanic ridge of which it forms a part are sometimes coated with ice clouds that may be the result of small outbursts.

Venus has several huge lava flows and volcanic complexes, with gently sloping sides marking them as shield volcanoes. The largest, Theia Mons, is a low mountain 510 miles (820 km) across, over four times the width of the Earth's largest shield volcano, Hawaii. No current volcanic activity has been traced on Venus.

Io, Jupiter's closest Galilean moon, has a thin atmosphere of sodium thrown out by active volcanoes, of which 11 are known. In the rarefied air, volcanic plumes leap up some 185 miles (300 km). One active volcano, Pele, sends up plumes that cover an area the size of Alaska.

Triton, Neptune's largest moon, produces a geyser-like eruption through its surface of frozen methane and nitrogen. Just 100 ft (30 m) down, the pressure is enough to liquefy nitrogen ice. The planet's surface cracks, releasing an explosion of liquid that turns to a gas plume and dumps vapour in a thin, dark layer on the surface.

GIANT VOLCANO *Olympus Mons on Mars, the largest volcano in the Solar System, formed gradually, probably from highly fluid lava.*

MOONS GALORE *Jupiter, by far the biggest planet, has 16 moons, including Io and Europa (left). Callisto (above), Jupiter's second-largest moon, has a rocky and icy surface riddled with craters, including one – Valhalla – that is 1600 miles (2600 km) across. It is one of six moons in the Solar System that are comparable in size to our own Moon (below). As yet, no one has identified any pattern in the Solar System's 61 known moons or their sizes.*

Moon (Earth)

Io (Jupiter)

Europa (Jupiter)

Ganymede (Jupiter)

Callisto (Jupiter)

Titan (Saturn)

Triton (Neptune)

years ago – for if it had happened earlier the watercourses would have been eroded away by dust storms. Given the rapid evaporation, the planet increasingly became a domain of volcanoes, frozen deserts and dust.

In brief, Mars advanced about the same distance as Venus down the road of planetary evolution, only to be killed off by its lack of an atmosphere. But at least it leaves open an intriguing suggestion. We have no idea how long the water flowed or exactly what happened to it. Perhaps a good proportion of it vanished underground again. Perhaps, in the lakes fed by the flowing rivers, some primitive life form evolved. Perhaps, like certain microorganisms on Earth, they evolved to live underground, away from the dying surface. Perhaps there, in small pockets of moisture, where Mars's internal heat melts the water, or frozen in suspended animation in the permafrost, they survive as living fossils left over from their planet's active past.

JUPITER, GIANT AMONG GIANTS

Beyond Mars, the character of the Solar System changes. A huge gap yawns – over 600 million miles (965 million km), six times the distance from the Earth to the Sun –

before the next planet, Jupiter. It and its companions, Saturn, Uranus and Neptune, are all gaseous giants with graceful rings of stones, dust and particles, and numerous moons. All have magnetic fields. Together, the four of them make an enlightening comparison with the development of the Earth, for all four are locked into an early stage of planetary development: all gas, with a very small core.

The prototype of these so-called Jovian planets is Jupiter itself, the largest body in the Solar System by far, 300 times the mass and 1000 times the volume of the Earth, $2^1/_2$ times more massive than all the other planets put together. But it is a great, soft ball consisting mainly of hydrogen and helium, a quarter the density of the Earth. Its atmosphere is torn by violent winds, probably induced by its fast spin – once every 9 hours 50 minutes at the equator. But at its poles its spin is five minutes slower, the result of its fluffy structure. The vast air currents circle the planet at different speeds, creating a series of bands and a hurricane-like, semi-permanent vortex known as the Great Red Spot, which itself is larger than the Earth.

As it plunged down through Jupiter's atmosphere, a spacecraft would find the hydrogen compressed by gravity to the point where the hydrogen liquefies. Farther down still, when the pressure mounts to 2 million times that of the Earth's atmosphere, the molecules of liquid hydrogen are broken down to form a very odd material, metallic hydrogen, which conducts electricity. Right in the centre of the planet there may be a hot core of heavy elements. The liquid metallic hydrogen, spinning fast and seething with convection currents induced by the hot core, acts as an electromagnet creating a powerful magnetic field ten times as strong as the Earth's.

In some ways Jupiter is a planet stuck in the earliest phase of formation, limited by its size and distance. Owing to its remoteness from the Sun, its gases remain cool and its mass binds them in place. So Jupiter's constituents have changed little since it was formed.

Of equal interest are Jupiter's moons. Four of them are large enough to be seen by a small telescope – in fact, Galileo discovered them and used them to support his view of the Solar System as a collection of small bodies orbiting a large central one. The comparison is apt, for two of them (Ganymede and Callisto) are larger than Mercury, while the other two (Io and Europa) are about the size of our Moon. A small fifth moon was discovered in 1892, and recent probes have discovered 11 even smaller ones – probably captured asteroids – making 16 in all.

The inner 'Galilean' satellites each have a character of their own. Because Io is close to its giant parent and its orbit is distorted by the other moons, it is continually being flexed and buckled, creating internal heat that escapes through volcanoes, producing a tenuous atmosphere of sodium. Europa is frozen, with an icy veneer, smooth as glass but cracked and ridged by shrinkage. The surface is quite recent because it covers any impact craters. Ganymede's surface, icy for aeons, is pockmarked by craters and patterned by cracks induced by crustal shifting long before it froze up. Callisto looks something like the Moon, but its impact craters are shallow, because it too is frozen and the surface does not fracture on impact. One

massive collision blasted a shallow basin 375 miles (600 km) across, throwing up waves of molten rock, which froze solid to form a series of concentric ripples, like a bull's-eye.

THE OUTER FRINGES

Moving towards the outer reaches of the Solar System, Saturn, Uranus and Neptune are smaller versions of Jupiter. They have hydrogen-rich atmospheres that become liquid, then metallic, at greater depths. The colours vary – Saturn is yellow and orange, Uranus green and Neptune blue, reflecting the varying amounts of their subsidiary gases, not their fundamental constituents. All have many moons – Saturn has 18, Uranus has five major ones and ten minor ones, Neptune has two major and at least six minor ones.

The really startling feature of these outer planets is Saturn's system of rings. These beautiful, regular, multicoloured rings, 170 000 miles (270 000 km) across and easily visible through an Earth-based telescope, were a mystery until the mid-19th century. Were they liquid or solid? In 1848, a French mathematician, Edouard Roche, proved that there is a limit within which any liquid or solid is disrupted by gravity. The rings were well within the 'Roche limit' for Saturn, and scientists agreed that they were probably made of particles of some sort.

Space probes have now spotted much less obvious rings around the other Jovian planets, and these features are better understood. Nevertheless, Saturn's rings still remain unique in their complexity. As

THE SECRET OF THE RINGS
From close up, Saturn's rings of dust and stones resolve into thousands of 'grooves' only $1/2$-1 mile (1-1.6 km) thick.

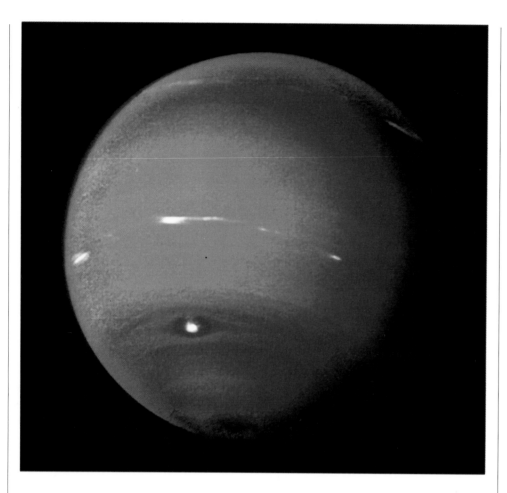

A PICTURE IN BLUE *Neptune has a blue atmosphere rich in methane. It also has rings, like Saturn, but they are too tenuous to be seen here.*

For almost 50 years after its discovery in 1930, it was nothing but a moving dot against the background of stars. Now more is known. It has a surface of frozen methane and, perhaps because part of its orbit brings it closer to the Sun, a diffuse methane atmosphere. And it is not alone. In 1978 something even smaller was found: a Plutonic moon, Charon, less than one-fifth of Pluto's mass, orbiting 12 500 miles (20 000 km) from its tiny parent.

Pluto is such an eccentric body, in many meanings of the word, that scientists suspect it is not a 'real' planet at all. Possibly it was once a moon of Neptune cast into outer darkness by a close encounter with Neptune's largest moon, Triton. Or perhaps it is the survivor of a collision that sent it on a ricochet into its present wild course.

We have now arrived at the outer reaches of the Solar System, from where the Sun is a mere pinpoint of light and the Earth completely invisible. Is that it? Is there anything beyond Pluto (or Neptune, as the case may be)? The answer, almost certainly, is yes. At least one small asteroid-like body has been found orbiting in the farther depths of space. Where there is one, there are probably more. And beyond lies the Oort Cloud with its reservoir of comets. However, it is unlikely that anything novel and large lies out there. If it was large, it would have shown up by now. It is fair to assume, therefore, that there is nothing more to be learned about the origins and fate of the Earth by exploring further. It is time to turn for home.

Roche suggested, they consist of tiny moons, millions of ice or ice-covered rocks ranging in size from lumps the size of golf balls to boulders as big as a house, all in regular orbit round Saturn's equator. There is not much to them: in total, the rings amount to a mass only one ten-billionth that of the Moon. Edge on, the rings virtually disappear, because they are only a few miles thick at most. Since the main rings stretch out over 435 000 miles (700 000 km), their orbital speeds vary, dividing them into three main bands, two of which subdivide into hundreds, perhaps thousands of ringlets so fine that at a distance they resemble grooves of a record. In fact, there is more to the rings than can be seen from Earth. Two spacecraft, Pioneer 11 and Voyager 1, have identified three other very faint ring systems stretching out many millions of miles.

No one knows how the rings formed. Most occupy a band so close to Saturn that no satellite could survive there. It would be torn apart by the planet's intense gravitational field. Perhaps they are the remains of one or more captured moons, or perhaps they are made of primordial particles that could never have formed moons.

Until January 21, 1979, the most distant member of the solar family, and its smallest planet, was thought to be Pluto, orbiting eccentrically almost 4 billion miles away. In fact, because of its eccentric orbit, it actually comes closer than Neptune for 20 years. Only in 1999 does it again become the farthest planet.

With one-fifth the mass of the Moon, Pluto is far smaller than many planetary satellites.

OUTERMOST INHABITANTS
Pluto, slightly smaller than our Moon, and its own moon, Charon, occupy the farthest reaches of the Solar System.

PARTNERSHIPS
IN SPACE

3

INFLUENTIAL PARTNER *The destiny of the Earth and its Moon are locked together.*

LIKE THE UNIVERSE ITSELF, THE EARTH AND ITS MOON CAN ONLY BE UNDERSTOOD WHEN THEIR ORIGINS ARE KNOWN, FOR THEIR STRUCTURES AND THE MECHANISMS THAT WERE ESTABLISHED IN THEIR EARLY STAGES — THEIR DISTANCE FROM THE SUN AND ORBIT AROUND IT, THEIR DISTANCE FROM EACH OTHER, THEIR GRAVITATIONAL FIELDS AND ROTATION SPEEDS — DICTATED HOW THEY BOTH EVOLVED. AND AS THE MOON CIRCLED THE EARTH, IT TUGGED AT ITS PARENT PLANET, AFFECTING EARTH'S TILT AND SPIN. ALL THESE FACTORS, TOGETHER WITH THE EARTH'S DRIFTING CONTINENTS, CHANGING ATMOSPHERE AND BOMBARDMENTS FROM SPACE, DEFINED THE CONDITIONS ON EARTH FROM WHICH LIFE WAS TO EMERGE.

OPPOSITES *The fertile Earth contrasts with the sterile Moon.*

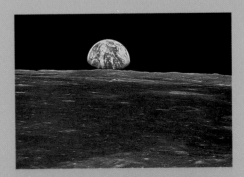

ELEMENTAL EARTH

Every culture has its creation myths, and until a century ago most Westerners accepted the version in the Book of Genesis. The mechanisms that shaped the early Earth, however, are far more fascinating than any creation myth.

Imagine a time before the planets evolved, when the Solar System took an entirely different form to now, consisting of a disc of loose debris circling confusedly, like a vastly more diffuse version of Saturn's rings. At several points larger lumps are forming, their differential motions (the differences in their speeds and direction) gradually sweeping clear, circular pathways. One of these lumps is some 100 million miles (1600 million km) from the Sun. It is the forming Earth.

The story of how it formed has emerged gradually during the course of the past two centuries, for it could only be understood once scientists understood the Earth's structure and the length of time that the process of formation occupied. Without an idea of the mechanisms involved and without a sense of time, there could be no understanding of the planet's defining traits – that unique combination of regularities and eccentricities that together created the preconditions for life.

Until the late 18th century, no one had any idea of how the Earth came into existence or how old it was. Before this time, Western scientific ideas were dominated by a combination of Christian theology and theories inherited from ancient Greece. All anyone knew was the Earth's size, quite accurately calculated by the Greek astronomer, Eratosthenes of Cyrene, some 2200 years ago. He calculated the Earth's circumference by measuring the Sun's position at the summer solstice from two different places, and this gave the Earth's volume and density.

As for Earth's origin, this was a blank. On one hand, the Universe could be seen as a divine machine, and eternal. But had it always existed? Not according to Genesis. The vague Biblical time scale was even specified with reassuring exactitude by an Irish archbishop, James Ussher, who in the early 17th century added up the generations mentioned in the Old Testament and worked out that the Earth was created in 4004 BC. By the mid-18th century, few but the most die-hard traditionalists really believed this, for it was becoming increasingly obvious that the processes of formation had taken a good deal longer.

In 1785, James Hutton, in his *Theory of the Earth*, proposed an opposite approach –

CREATING THE SUN AND STARS
*A 16th-century French stained-glass window records God ordaining the emergence of the Universe from chaos.
Right: By recording the Sun's position from two places at the summer solstice, Eratosthenes calculated the circumference of the Earth to be 26 000 miles (42 000 km).*

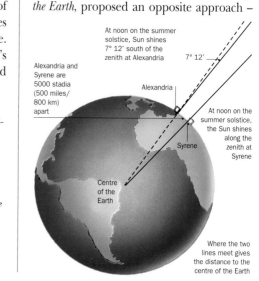

At noon on the summer solstice, Sun shines 7° 12' south of the zenith at Alexandria   7° 12'

Alexandria and Syrene are 5000 stadia (500 miles/ 800 km) apart

Alexandria

At noon on the summer solstice, the Sun shines along the zenith at Syrene

Syrene

Centre of the Earth

Where the two lines meet gives the distance to the centre of the Earth

that all the forces that formed the Earth were still in operation, and that the Biblical account of Creation was a figment of the imagination, at least in its details. In his view, the Earth recycled itself ad infinitum by erosion and uplift, bracketed only by divine intervention to start and stop it. He never guessed at a time scale because 'time . . . is to nature endless and as nothing'. In any event, the Earth was older than Ussher had imagined.

Debate about the origin and age of the Earth rumbled on through the 19th century, increasing in complexity as the sciences of geology, physics and biology developed. At heart, the controversy involved a clash of principles: did the Earth and its life forms develop gradually and uniformly over an extended time scale, with the forces that currently exist, or was there a time when catastrophes forced the pace?

The greatest catastrophe would have been the Biblical Flood. In the early 19th century, it was not seriously doubted that the Flood had actually happened. Many scientists shared with theologians the assumption that there had once been a universal deluge. In 1823, Britain's leading geologist, William Buckland, in his *Relics of the Flood*, based his argument on sediments and bones found in Yorkshire caves. His

NATURAL PROCESS Melting glaciers, rather than a universal deluge, explains the presence of floodwater deposits in northern Europe.

finds, he said, were conclusive. 'The grand fact of a universal deluge at no very remote period is proved on grounds so decisive and incontrovertible, that had we never heard of such an event from Scripture or any other authority, Geology of itself must have called in the assistance of some such catastrophe.'

Within 20 years, his argument had collapsed, because other 'Flood-water' deposits in England and elsewhere proved to be of a different age. Nor were they universal: some places farther south in Europe had had no floods. Buckland had found evidence not of one flood, but many. They were not universal, either, but involved only northern Europe, and were caused by water from melting glaciers at the end of the last Ice Age. At the time, no one would ever have thought that northern Europe had quite recently been covered by an ice sheet.

In 1830, Sir Charles Lyell set out Hutton's case with much greater force, implying that since the past was pretty much the same

EXTENDING THE PAST *Charles Lyell, the doyen of 19th-century geologists, insisted that the Earth had been formed gradually by natural processes.*

as the present in geological terms, the past must extend indefinitely – not just millions, but thousands of millions of years back.

This view was denied by the physicist Lord Kelvin, who dismissed the idea that the Earth could be that old. He based his argument on the fact that the Earth was losing heat – a prodigious amount over the whole surface of the Earth. Kelvin argued that if heat had been produced in this quantity over billions of years, it would have melted the whole Earth 100 times over. Supposing it had radiated its heat regularly, the Earth could only be 400 million years old at most, probably no more than 100 million years, a figure that he cited during a 40 year campaign to have his theory recognised. Perhaps, he

suggested, 20 million might turn out to be the right figure.

But if this was so, how could there have been enough time for a gradual evolution of animal life, as Charles Darwin proposed in his *Origin of Species*. Darwin himself argued that species evolved by amassing slow and steady changes over a time scale he never defined. Evolutionists conceded that they had a problem. In 1870, one of the leading evolutionists, Alfred Russel Wallace, said he could imagine all evolution being squeezed into 24 million years.

The controversy ended quite suddenly in 1903, when the French physicist Pierre Curie announced that radium salts (a rare radioactive metallic material) constantly released newly generated heat. The discovery of this process, radioactivity, revealed the Earth's inner source of heat and at a stroke invalidated Kelvin's theory. Within a decade, scientists were talking of Earth's history in terms of billions of years, and had embraced the idea that the Earth had a progressive biography from youthful plasticity to today's middle-aged stability.

PROBING THE DEPTHS

Information about the interior of the Earth is gained by, in effect, 'X-raying' it with seismic waves. These are the vibrations

THE MAN WHO DATED CREATION

In the mid-17th century, no one had any idea of the true age of the Earth. Though some other cultures (such as Hinduism) had long relied on vastly extended time scales, Western tradition had little to go on but the Bible, which was almost universally accepted by European thinkers as the literal Word of God. With the Bible as the only source of information, it was easy to estimate the age of the Earth from the generations descended from Adam listed in the Old Testament.

In 1620, an Irish archbishop, James Ussher (1581-1656), went farther: in his *Sacred Chronology*, he added up the ages of Adam's descendants, and came up with a reassuringly exact date. The world had been created, he said, in 4004 BC. His calculations were further refined by Dr John Lightfoot,

OUTDATED *Archbishop Ussher devised a chronology of world history that depended on Biblical genealogies.*

Vice-Chancellor of Cambridge University, who declared the moment of Creation to be Sunday, October 23, 4004 BC, at 9 am. This chronology, printed in Bibles for the next two centuries, stood as the

'truth' until new geological discoveries made it obviously ridiculous.

Ridicule, however, was no just reward for Archbishop Ussher's intellectual ability and achievements. As a professor of Trinity College Dublin at the age of 26, he was a man of great scholarship with a formidable knowledge of early Church history, an expert in Semitic languages and a collector of Oriental manuscripts. He was widely admired for his detailed work on the letters of the 2nd-century martyr, Ignatius, and as Archbishop of Armagh was deeply involved in Anglo-Irish politics until the outbreak of the English Civil War in 1640 forced him to remain in England. He died in 1656, never knowing that his undoubted erudition would make him a byword for naivety and arrogance.

BREAKTHROUGH *Pierre Curie, together with his wife Marie, discovered radioactivity, the process that provides the Earth's core with its energy.*

produced by earthquakes and artificial explosions, both of which make the Earth ring like a bell. Seismic waves are of four types. Two travel near the surface and reveal nothing of the interior. The other two types, called primary (P) and secondary (S) waves, travel deep inside the Earth and can be detected at great distances from the earthquakes that produce them. A P-wave is like a sound wave in air, and its vibrations travel in the direction of motion, while an S-wave sets up a sideways vibration. P-waves are the faster of the two, and travel through both solids and liquids; S-waves are slower, and travel through solids only.

Both types travel in a straight line through homogeneous material. Where they encounter a change in density they are deflected and the direction of their path changes. By timing the arrival of waves at different

P AND S WAVES *Seismic waves can be used to discover the internal structure of the Earth. An S-wave (right) creates a ripple effect but only through solids. A P-wave (above right) creates a shunting movement and can travel through solids and liquids.*

points, and measuring how far they have been deflected, scientists can derive much information about the materials through which the waves have passed The most important discovery was that the Earth's density changes suddenly, indicating the existence of shells.

At Earth's centre is the core, identified because of a puzzle that emerged in the late 19th century. When large earthquakes were recorded, some places on the Earth's surface received no vibrations at all. In 1906, these 'dead spots' were explained. Something in the centre of the Earth created a shadow zone that appeared to deflect any vibrations reaching it. That 'something' was a core with a different structure, but there was no telling yet whether it was solid or liquid, or how deep it lay.

In 1914, the German physicist Beno Gutenberg showed that S-waves do not penetrate beyond about 1800 miles (2900 km). And as S-waves do not travel through liquids, this meant that at that depth – about half way to the centre of the Earth – the core was

EARTHQUAKE MIMIC *A 26 ton Vibroseis truck transmits vibrations into the Earth that reveal rock structure and predict earthquakes.*

liquid. In 1936 a Danish seismologist, Inge Lehmann, found that P-waves do show up in the 'shadow zone' after all, but faintly and very much delayed. The best explanation was that only the outer part of the core was liquid and that there is a solid inner core,

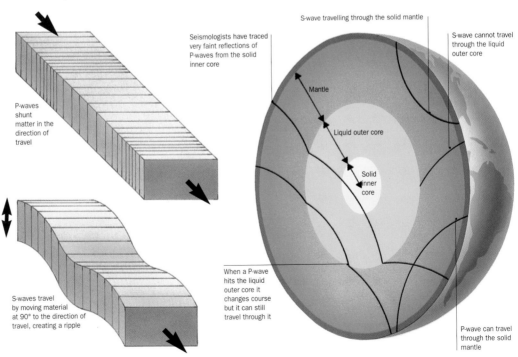

P-waves shunt matter in the direction of travel

S-waves travel by moving material at 90° to the direction of travel, creating a ripple

Seismologists have traced very faint reflections of P-waves from the solid inner core

When a P-wave hits the liquid outer core it changes course but it can still travel through it

S-wave travelling through the solid mantle

S-wave cannot travel through the liquid outer core

Mantle

Liquid outer core

Solid inner core

P-wave can travel through the solid mantle

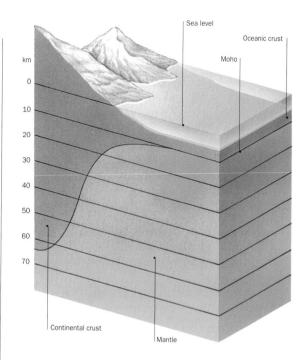

Sea level

Oceanic crust

Moho

km

0

10

20

30

40

50

60

70

Continental crust

Mantle

VARIABLE DEPTH *The Earth's crust can project as far down as 44 miles (70 km) beneath mountain ranges. The oceanic crust is thin by comparison.*

1500 miles (2500 km) across. Since then, working with more refined seismographs, geologists have recorded vibrations bouncing off the boundary of the solid inner core.

The core is dense – 13 times denser than water. This is about the density of iron at those depths and pressures. Apparently, since the fires in the interior separate out the lighter from the heavier materials, the two parts of the core consist mainly of iron, probably mixed with some nickel, rising to a temperature of about 6000°C (11 000°F) at the centre. This would be more than enough to melt the iron-nickel alloy, were it not for the pressure of the outer layers. It is this pressure that turns the inner core solid.

Over the core is the so-called mantle, 1800 miles (2900 km) thick. In the 1920s, Beno Gutenberg showed that this has a three-part structure. Deep down, it seemed to be rigid. Then – to his surprise – the consistency began to change. The waves moved faster. Near the top, down to a depth of about 130 miles (200 km), they moved so fast that the material they were moving through could not be a proper solid. The only explanation was that the rock at that level is so hot that it acts as a very turgid liquid, a sort of geological treacle.

The outer shell is the crust, which was discovered in 1909 by a Croatian geologist, Andrija Mohorovicic, who was researching earthquakes in the Balkans when he discovered that the P and S waves suddenly increased in speed a few miles below the surface. Later it was discovered that this was a worldwide phenomenon, with the change taking place at a depth that varies from 19-25 miles (30-40 km) under the continents to 6-7 miles (9.5-11 km) beneath the oceans. On top, the rocks are mainly granite, riding like a raft on a lower, heavier crust of basalt. The discovery won an otherwise unknown scientist a place in the scientific pantheon: the boundary between the upper mantle and the crust is named after him: the Mohorovicic discontinuity, or Moho for short.

THE LIVING CORE

Now it is possible to return to the image of the forming Earth. Slowly, the growing body began to acquire its warm heart, a core hot enough to melt rock. How did this happen? Partly through the accretion, or fusing together, of particles, which provided a certain warmth, and partly through the Sun's heat, which provided enough heat to raise the temperature to some 20 000°C (36 000°F). However, this source of energy lay on the surface of the planet, and would soon have radiated away, leaving a subsurface temperature of no more than about 200°C (390°F). To heat the interior, another source of energy

REVEALING THEIR AGE *To date the world's oldest rocks, such as those on Ellesmere Island in the Canadian Arctic (right), scientists measure the decay in radioactivity in uranium-rich samples. Below: The energy being emitted by uranium oxide is not obvious in ordinary light (left) but can be seen under fluorescent light (right).*

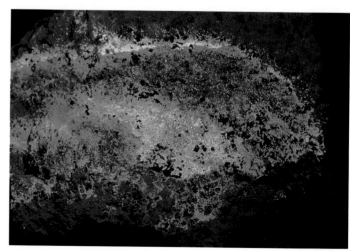

was required; this, discovered by Curie, was radioactive decay.

Radioactive decay is the result of spontaneous disintegration of an unstable nucleus, by shedding one or more particles, to form a stable nucleus. The new element created by this process may or may not be radioactive itself. The time taken for half the atoms to transform themselves into the new element is called the half-life, and differs from one element to another. For example, uranium 238 decays to lead 206 and has a half-life of 4.5 billion years; carbon 14 decays to nitrogen 14 and has a half-life of 5730 years.

The radioactive elements responsible for heating the interior of the Earth are of two types: short-lived elements with half-lives of 1-10 million years; and long-lived elements with half-lives of 1 billion years. Very few of the short-lived ones now survive, but there are enough of the long-lived elements – uranium, thorium, potassium – to sustain the reactions that produce decay.

As the young Earth grew in size with every new collision, the growing layers of

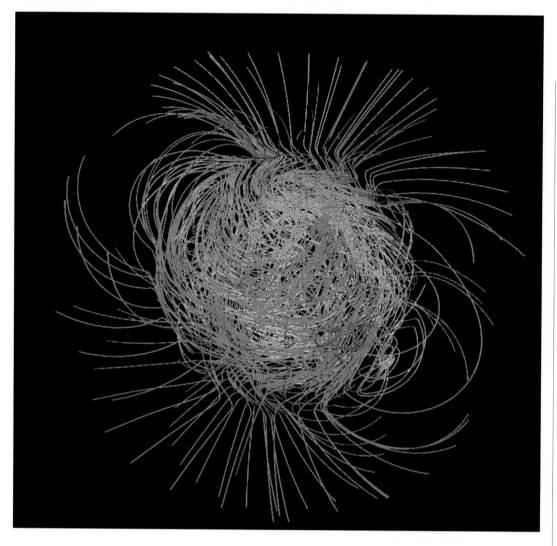

rock blanketed the interior, preventing heat at the core from escaping. Steadily, over tens of millions of years, the centre began to heat up. The effect was comparable to that of a flame beneath a pan. Once the Earth reached a critical size, the temperature rose high enough – some 6000°C (11 000°F) – to melt rock. Denser materials such as iron sank to the core, and lighter ones rose to the surface. Lighter gases, like the hydrogen and helium that formed such a high percentage of the original material, were mostly blown away by the heat. A new atmosphere, consisting mainly of carbon dioxide, drifted out from the slowly churning rock.

As if in a planetary steelworks, slag rose to the Earth's surface, forming a rich mixture of oxygen (which is by far the commonest element, forming

DRIFTING *Compass readings show the drift in magnetic north. The magnetic field reversed 750 000 years ago.*

N 750 000 years ago

N
1580 11° East

N
1660 Due North

N
1820 24° West

N
1970 7° West

MAGNETIC MODEL *The orange lines indicate where the north pole of a magnet is attracted to Earth; the blue lines show where it is repelled. Earth's magnetic field acts like a bar magnet buried at the centre of the planet (below).*

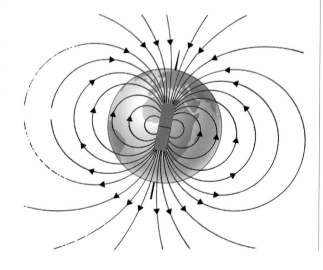

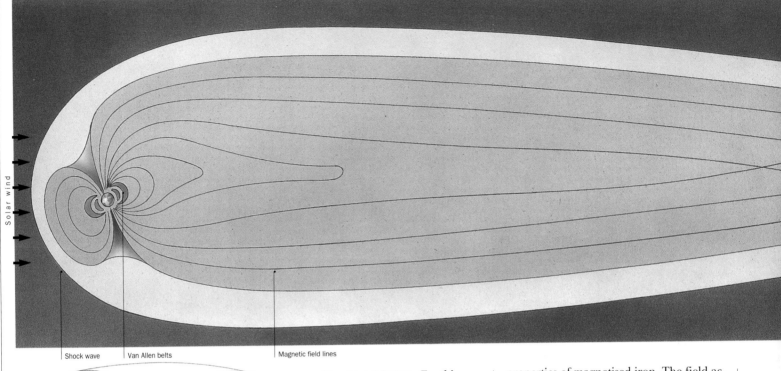

Shock wave   Van Allen belts   Magnetic field lines

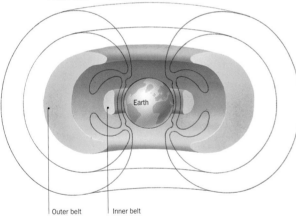

Outer belt   Inner belt

*VAN ALLEN BELTS   Earth's magnetic field creates a cavity in the solar wind. A shock wave marks the leading edge; most of the particles in the wind are deflected around the Earth.
Left: Earth's magnetism traps charged particles in two rings, known as the Van Allen belts, around the Earth at the Equator.*

just over 46 per cent of all Earth-matter) and silicon (about 28 per cent), together with the other most common elements: aluminium, iron, calcium, potassium, sodium, magnesium and titanium. Together, these elements accounted for over 99 per cent of the material that separated out of the core. All the other 81 natural elements – including the prime source of radioactivity, uranium – made up just 0.3 per cent.

The separation into core, mantle and crust occurred very early in the Earth's history. Since radioactive elements decay at a known speed, the proportion of these elements in ancient rocks reveals their age.

The oldest rocks that have been studied in North America – in Western Greenland and Canada's North-west Territories – are about 4000 million years old. The oldest in Australia are found in the western half of the continent, and were formed as long as 4200 million years ago. By working out how fast the primitive Earth heated, cooked up its crust and cooled enough for the crust to harden, scientists estimate that the Earth itself is about 4600 million years old.

The iron core is responsible for another of Earth's peculiarities: its magnetic field, signs of which were first recorded around 500 BC, when Thales of Miletus noted the

properties of magnetised iron. The field acquired a practical significance with the invention of the compass, some time after 1000 AD. Its effect is similar to that produced by a bar magnet or a magnetised sphere, as William Gilbert, physician to Queen Elizabeth I, pointed out. The magnetic field has two poles that are close to, but not exactly at, the Earth's geographical poles. The magnetic North Pole is in Canada, 12° and 812 miles (1306 km) from the geographical North Pole. The magnetic South Pole is in Antarctica, 25.6° and 1775 miles (2858 km) from the geographical South Pole.

The magnetic field is not stable. Scientists have been measuring the field and its changes for some 400 years, discovering in the process that it shifts around quite fast. This means that the cause of the shifts cannot have anything to do with the Earth's crust. Besides, there are irregularities in the field that do not correlate with surface features. Rather, it is the shifting liquid core that creates the field.

The magnetic field has another feature, discovered in the early 20th century when scientists began to examine sedimentary rocks that are naturally magnetic. The magnetism in liquid rock aligns itself with Earth's magnetic field and retains this magnetic orientation after it has cooled down. These rocks therefore provide a record of the direction of Earth's magnetic field at the time they were formed. Scientists discovered that rocks on the ocean floor alternated between normal and reversed polarities. By dating the rocks, they could chart a record of changes in the magnetic field. These studies show that the Earth's magnetic field completely reverses itself, on a very rough average, about once every 400 000 years. During the past 71 million years there have been 171 such field reversals, the most recent of which was 750 000 years ago. Why they should occur is at present a complete mystery, but sometime soon, it seems, we are due for another one, with consequences that can only be guessed at.

Such irregularity stems from the movement of the liquid outer core where the magnetic field is generated. However, permanent magnetism is not thought to be possible in a liquid, because the motion would destroy it. Another way of producing a magnetic field is with an electric current. Being metallic, the Earth's core is a good conductor of electricity, but any electric current started in the core would die away over tens of thousands of years. The alternative is that electricity is constantly being generated.

The Earth is, in effect, a dynamo and an electromagnet. In an artificial dynamo, an electric conductor driven round in a magnetic field produces a flow of electricity. In a so-called self-exciting dynamo, part of the electric current can be used to add to the magnetic field, which in its turn feeds the electric current. Artificial dynamos that do this, like those in many power stations, are extremely complex. Though no one knows the details, scientists suppose that the Earth's liquid core contains complex movements created by convection that mirror the structure of a self-exciting dynamo.

Whatever its cause, the Earth's magnetic field stretches way out into space, where it assumes a peculiar shape under the influence of the solar wind, the stream of

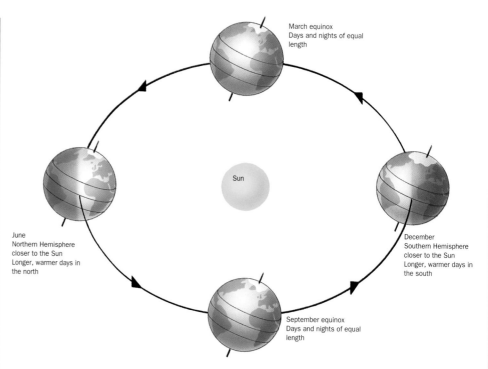

March equinox
Days and nights of equal length

June
Northern Hemisphere closer to the Sun
Longer, warmer days in the north

Sun

December
Southern Hemisphere closer to the Sun
Longer, warmer days in the south

September equinox
Days and nights of equal length

charged particles emitted by the Sun. The Earth's magnetic field is flattened on the side facing the Sun, forming a shock wave some 43 000 miles (69 000 km) out from Earth. On the opposite side, the field stretches out in a tail more than 7.5 million miles (12.5 million km) long, reaching beyond the Moon's orbit.

The Earth's magnetic field strongly affects the charged particles given off by the Sun. Low-energy particles are pushed clear; some higher-energy particles are trapped in two doughnut-shaped bands of radiation above the Equator, named the Van Allen belts in 1958 after their American discoverer, James Van Allen of the University of Iowa. Other higher-energy particles are deflected along the magnetic lines of force towards the poles. There, they produce a curtain of coloured, waving lights known as an aurora. The word derives from the Latin for 'dawn', but the lights have nothing to do with direct sunlight, or reflected sunlight.

As the particles enter the upper atmosphere, they interact with high-flying atoms as if in a giant neon light, creating glowing drapery some 3000 miles (4800 km) long, over 100 miles (160 km) high, petering out about 65 miles (105 km) above the

TILTING PLANET *The tilt of the Earth's axis means that days and nights are unequal in length except at the spring and autumn equinoxes.*

surface of the Earth. These polar lights – or *aurora polaris* in their Latin form – are common in high latitudes, where they are known as northern and southern lights (aurora borealis and aurora australis). They are sometimes seen away from the poles. The aurora borealis have occasionally been seen as far south as the Mediterranean, and the aurora australis was sighted at Apia in Western Samoa in 1921. Once, in 1957-9, a time of numerous magnetic disturbances on the Sun, auroras were seen in Mexico.

UNIQUE TRAITS OF THE EARTH IN SPACE

When the Earth formed, various peculiarities were established that would have crucial effects once it cooled enough to evolve climates suitable for life. These oddities concerned its shape, its attitude or tilt, its spin and the path of its orbit around the Sun, which in combination would affect the amount of heat received from the Sun. The differences are minute, but they add up. As the Earth cooled to establish a precarious balance between liquid water and ice, the differences became large enough to affect that balance and contribute to the onset and end of ice ages.

The first of the Earth's eccentricities involves the tilt of the Earth as it orbits the Sun. In an ideal solar system, all the planets would be spinning 'upright'. As we have seen, they are not, probably as the result of the random nature of the early impacts that built them up. In any event, by the time the Earth was established as a sphere, it was spinning about 23.5 per cent from the vertical. In the course of its annual swing around the Sun, the Earth tilts first the Northern Hemisphere and then the Southern towards the Sun. Almost all the time, therefore, days and nights are of unequal length everywhere.

Only when the Sun crosses the Equator, once every six months, is there an 'equinox', when the day and night are of equal length. The effects of this must have made little difference in the early stages of the Earth's formation, but it introduced fundamental elements into the mechanisms controlling the climate and weather.

A second peculiarity concerned the shape of the Earth. Usually, it is called a sphere, a ball, a globe – words that suggest that our planet is uniformly round. It is not; or not quite, a fact that was first discovered in the late 17th century. During 1671-3, an otherwise unknown astronomer named John Richer, on a French expedition to Cayenne in French

SPHERE OF ACTIVITY *Seen from space, the Earth's marbling of sea, land and clouds is evidence of her dynamism and extended evolution.*

Guiana, South America, discovered that a pendulum there swung more slowly by a minute amount than in Paris.

According to Newton's laws, which were published 15 years later, there was only one possible explanation. The Cayenne pendulum was farther from the centre of the Earth than the one in Paris. This would have been the case, as Newton himself realised, because Cayenne is nearer the Equator than is Paris, and the Earth is fatter at the Equator due to the effects of the centrifugal force created by the Earth's spin. The difference between the polar and equatorial circumference is tiny, a mere 35 miles (55 km), but it is nevertheless of immense significance.

Other peculiarities about the Earth's spin and orbit were first analysed by a Serbian engineer and geophysicist, Milutin Milankovitch, in the 1930s, but they were recognised as valid only in the 1970s. Milankovitch's theory involved three separate eccentricities that together combine to reinforce each other. The first and longest cycle lasts just less than 100 000 years, during which the Earth's orbit stretches from a near-circle into more of an ellipse as the result of the gravitational influence of the other planets.

An understanding of the second and third cycles discovered by Milankovitch demands a closer look at another member of the Solar System, without which the Earth would not exist in its present form: the Moon.

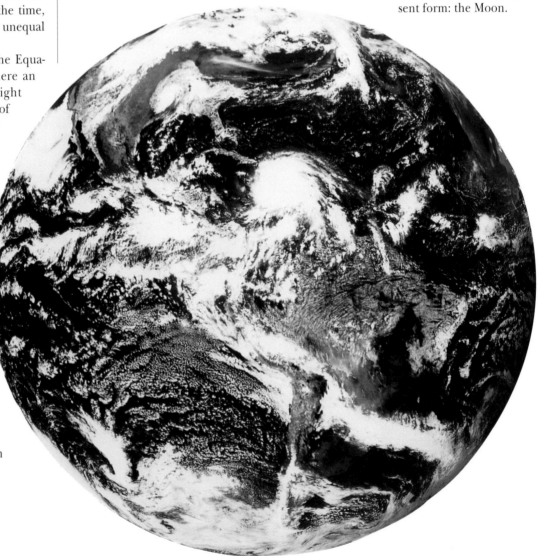

SISTER MOON

The Moon was a dead world before even the first living cells appeared on Earth. It is, in effect, a geological fossil, preserved by its lack of an atmosphere and geological activity, but rich in clues to the Earth's own past.

The Moon is a peculiar object. It is so close to us – just 250 000 miles (400 000 km), a distance surpassed by many an old Volkswagen – and so familiar that its peculiarities are easily forgotten. Measuring 2160 miles (3476 km) across, it is by far the largest satellite in the Solar System in proportion to its mother planet. There are bigger satellites circling Jupiter and Saturn, but in proportion to the size of their parent planets they are tiny, and their influence on their parent planets is tens of thousands of times less powerful.

Although close, the Moon is nonetheless mysterious. In one sense, it has always been easy to observe. Its largest features, the huge, dark 'seas' as they were called, are visible to the naked eye, and even a small pair of binoculars reveals dozens of craters. The ancient Greeks had a rough idea of the Moon's size from the size of its shadow during eclipses. They also knew that its orbit was a little off-centre (its distance from the Earth varies by up to 35 000 miles/56 000 km), that it circled the Earth roughly, but not exactly, on the same plane as the Earth circled the Sun, and that there were other minor variations, which were explained only in the 17th century as being caused by the gravitational pull of the Sun. It has always been apparent that the Moon points only one side towards the Earth, because the time it takes to orbit the Earth and the time it takes to spin once on its axis are the same – 27.3 days (though because of the Earth's motion round the Sun, the Moon takes another 2.2 days to return to the same position relative to the Sun).

The Moon's orbit is extremely complex in detail. One of its eccentricities, a rocking motion, ensures that year on year, it rises and sets at a slightly different place, in a pattern that is repeated every 18$\frac{1}{2}$

THE DARK SIDE *From space it is possible to see the hidden side of the Moon. It is as cratered as the side facing Earth, but contains fewer dark 'seas'.*

the inch. It cannot, and for two centuries scientists wrestled with what came to be called the 'three bodies problem' before discovering that the equations had no solution. In the early 20th century, the tables from which its position could be derived ran to as many as 650 pages. Someone would have to work full time at the tables to extract the data fast enough just to keep up with the changing position of the Moon!

In other ways, too, the Moon remained an enigmatic body. One reason for this was because the far side of the Moon was unknowable. Another was that no Earth-based telescope can penetrate Earth's turbulent atmosphere well enough to resolve lunar features much smaller than 500 yd (450 m) across. Craters, mountains, escarpments and rifts were identified, however, and astronomers gave these features romantic names in a tradition established by Galileo in the 17th century – Mare Imbrium (Sea of Showers), Mare Tranquilitatis (Sea of Tranquillity), Sinus Iridium (Bay of Rainbows). Although the tradition of using Latin names was later broken, these large, long-recognised features retain theirs.

Yet clearly the surface was so different from that of Earth that scientists had no clue as to the Moon's age or how the craters were formed. Were they volcanic? Or the result of meteorite impacts? Or a combination? Or were they caused by something else entirely? One strange suggestion was that the mountains were coral atolls, another that the craters were the result of nuclear war fought by a race who were, quite literally, lunatics. And why the peculiar synchronicity of the Moon's spin and orbit? Only the coming of the Space Age made answers to these questions possible and threw more light on Earth's own history in the process.

Meanwhile, astronomers did their best with the information available. The 18th-century astronomer Sir William Herschel believed the Moon to be inhabited, and that it therefore had an atmosphere. But by the mid-20th century, scientists agreed that the Moon had no air, that the seas were not

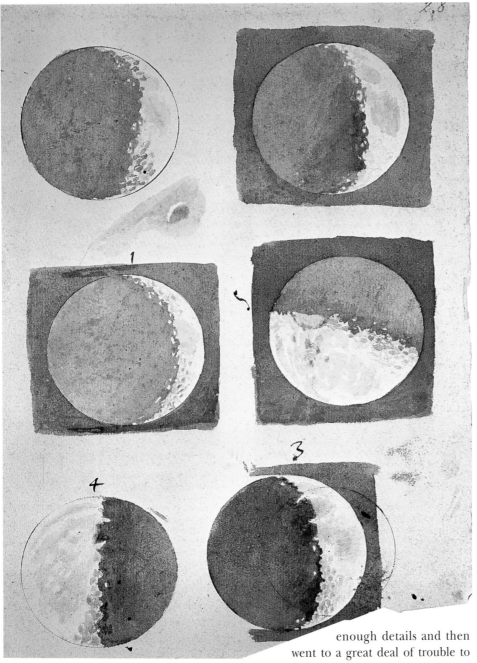

MOON DRAWINGS *Galileo's telescope observations, recorded in his drawings, revealed the Moon's surface to be cratered and mountainous.*

years. This and other regularities obsessed earlier cultures, which depended on accurate predictions of the position of the Moon for agricultural and religious purposes. The builders of Stonehenge, for instance, took generations to build up enough details and then went to a great deal of trouble to record them in stone, marking the extreme positions of moonrise and moonset (among many other things) over the 18½ year cycle.

LUNAR ENIGMAS

After Sir Isaac Newton had formulated the laws of planetary motion in a usable form in the 17th century, scientists assumed that the Moon's motion, defined by the forces linking the Earth, Moon and Sun, could be pinned down mathematically to watery, and that it was a dead world (or almost dead, for occasional minute and transient glows were seen, the result of minor emissions of gas and dust).

THE APOLLO BREAKTHROUGH

From pictures sent back from lunar probes, orbiters and landers – the first ones being Russian – scientists in the mid-1960s thought the Moon's surface was solid enough to land on. But no one was absolutely certain until the USA overtook Russia in the space race in the late 1960s. US orbiters and landers mapped the surface, but left some final, practical questions unanswered. When Neil Armstrong and Edwin 'Buzz' Aldrin dropped down towards the lunar surface in July 1969, there was a niggling concern that their spidery *Eagle* lander would vanish into dust. In the event, they had an opposite problem. Armstrong had to guide the craft away from a mass of boulders before setting down on the powdery but hard surface on

CALCULATING DISTANCE Laser reflectors placed on the Moon by Apollo astronauts enabled scientists to measure the Moon's distance to within 1 in (2.5 cm).

which he could take his 'small step', mankind's 'giant leap'.

The Apollo missions led to an enormous increase in our understanding of the Moon. By placing laser reflectors on the Moon, US astronauts were able to determine the Moon's position to within 1 in (2.5 cm). Probably the most important discovery was the age of the Moon from the 800 lb (360 kg) of rock brought back by the astronauts. By examining the decay in the rocks' radioactive elements, scientists showed that the youngest rocks on the Moon were between 3.9 and 3.2 billion years old. This was extraordinary, because in many ways the surface looks new-minted with outflows of lava. It shows that the

Moon was in effect geologically dead (that is, its geological structure had stopped changing) 2.6 billion years before life emerged on Earth.

Studies of the Moon rocks also revealed that all the craters on the Moon were the result of meteorite impacts. Some of the more ancient craters – the 'seas' – are truly enormous, produced by bodies miles across travelling at tens of miles per second. Such impacts liquefied the Moon's rock and caused floods of lava, which flowed out to form the huge circular basins. In a sense, they are seas, but seas of frozen rock. These areas, and many of the other larger craters, are cut into by smaller, later impacts.

What is the Moon made of? For decades, it had been known that the Moon was rocky. Its mass was known from its effect on Earth's tides, and so was its density, which is slightly less than the Earth's. It turned out that the lava filling the 'seas' was similar to terrestrial lava, and that it had a mantle of lighter-weight silicates (a common mineral). This indicates that in its youth the Moon generated enough heat to melt rock. However, it has only a very weak magnetic field, $1/10\,000$ that of Earth's, and

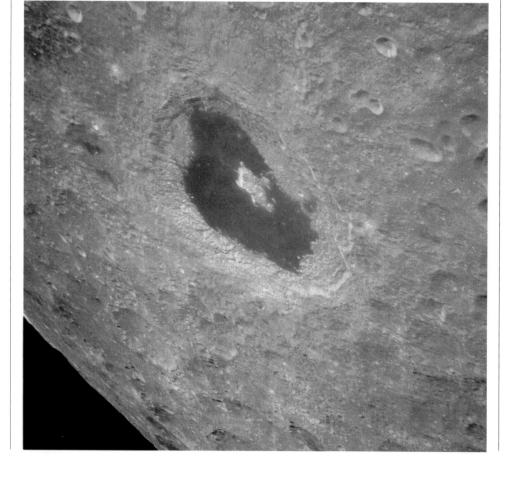

IMPACT CRATER The crater Tsiolkovsky is one of the four 'seas' on the far side of the Moon. It is 95 miles (150 km) across and has a central peak.

AT WORK ON THE MOON

Edwin 'Buzz' Aldrin, who with Neil Armstrong made up the first pair of US astronauts to walk the lunar surface on July 20, 1969, gave this description of his experience:

'The Moon was a very natural and very pleasant environment in which to work. It had many of the advantages of zero-gravity, but it was in a sense less lonesome than zero-G, where you always have to pay attention to securing attachment points to give you some means of leverage. In one-sixth gravity, on the Moon, you had a distinct feeling of being somewhere, and you had a constant, though at many times ill-defined, sense of direction and force.

'As we deployed our experiments on the surface we had to jettison things like lanyards, retaining fasteners, etc, and some of these we tossed away. The objects would go away with a slow, lazy motion.

'Odor is very subjective, but to me there was a distinct smell to the lunar material – pungent, like gunpowder or spent cap-pistol caps. We carted a fair amount of lunar dust back inside the vehicle with us, either on our suits and boots or on the conveyor system we used to get boxes and equipment back inside. We did notice the odor right away.'

LUNAR SAMPLES *An Apollo astronaut leaves the Lunar Rover to gather samples of lunar soil for analysis.*

---

therefore probably has no liquid core (essential to the creation of electromagnetism).

Space-age research also raised new and intriguing problems. The far side of the Moon turned out to be rather different from the near side. It has fewer large craters, and only four 'seas'. Clearly, any theory as to why this is so involved both the Moon's development and its interaction with its parent, the Earth. When looked at on this scale and over this length of time, the Moon turned out not to be dead at all; on the contrary, through its gravitational influence it has been a lively contributor to the evolution of the Earth.

THE MOON'S ORIGINS

Despite all the advances in technology and knowledge, there is still no wholly satisfying theory that accounts for the Moon's creation. Some scientists have favoured an accretion theory that sees the Moon as a smaller version of the Earth, gradually growing as lumps of matter fused together. Others have proposed that the Moon was a small planet captured by the Earth's gravity.

FIRST STEPS

The first astronauts to visit the Moon left a commemorative plaque on the surface: 'Here men from the planet Earth first set foot upon the Moon, July 1969 AD. We came in peace for all mankind.'

Of the various theories that have been proposed, however, one scenario has now emerged as the most acceptable.

This scenario, which has been devised with computer simulations, sees the Moon as a product of the young Earth, the result of a massive collision with an asteroid; an asteroid, for example, the size of Mars. About 4.6 billion years ago, the asteroid struck the Earth a glancing blow, destroying itself, blasting away much of the Earth's primitive mantle and some of its core, vaporising volatile gases, and creating a huge amount of liquefied debris, much of which fell together to form the primitive Moon. All this took less than a day, leaving an ill-formed globe circling the Earth. Under the influence of the Sun's gravity, the early Moon fell into an orbit roughly on a level with the Earth's path round the Sun, and roughly on a level with the Earth's spin.

As the Moon formed, spinning fast as it did so, the Earth generated tidal bulges in the liquid rock. The bulges were constantly pulled towards the Earth, creating

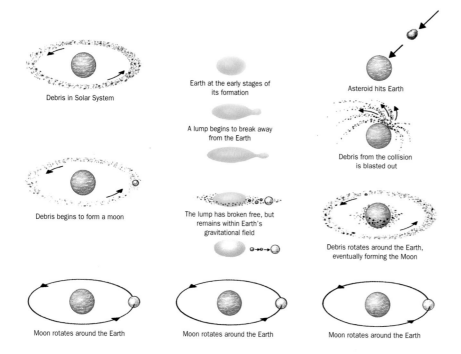

Debris in Solar System

Earth at the early stages of its formation

Asteroid hits Earth

Debris begins to form a moon

A lump begins to break away from the Earth

Debris from the collision is blasted out

The lump has broken free, but remains within Earth's gravitational field

Debris rotates around the Earth, eventually forming the Moon

Moon rotates around the Earth

Moon rotates around the Earth

Moon rotates around the Earth

CONFLICTING THEORIES *The Moon may have formed from the accretion of debris in the Solar System (left), from a lump breaking off the forming Earth (centre), or from debris from an asteroid impact (right).*

friction, which dragged at the Moon, slowing it until it steadied with one face towards its parent.

For perhaps 200 million years, the debris continued to fall on the Moon, keeping the surface molten, so that Moon and Earth formed in parallel, with heavier elements sinking inwards, gathering radioactive material at the core. A magnetic field built up. Over the next 500 million years, the Moon's crust hardened, bombarded all the time by debris left over both from its own formation and the formation of the Solar System. Under these impacts and still-active geological forces, such as volcanoes and earthquakes, on the Moon itself, mountains formed and lava filled the huge impact craters, creating the 'seas'.

Finally by about 3 billion years ago, the bombardment of debris died away and impacts became rarer. The Moon's interior was not hot enough on its own to generate an active crust. The core began to cool, and

the magnetic field died away. The surface locked into senescence – becoming so hard and rigid that it could no longer repair itself after an impact; and the asteroids that continued to strike – smaller and fewer in number – permanently scarred the surface. From then until now, the only other active force was a rain of minor debris, the interplanetary dust particles that on Earth appear as shooting stars. On the Moon they rain down directly onto the surface, eroding the rock away to form a light dust.

The theory explains a good deal: why the Moon is similar to, but different from the Earth; why there are several 'seas' on the Earth-facing side and only four on the hidden side; why the early Moon was molten; why it has a core; why

INSDIE THE MOON *The Moon probably has a crust, mantle and core. Its craters were formed by meteor impacts. Material blasted out by the impacts fell back in streaks, forming ray-like patterns.*

it has no volatile gases; why it presents only one face to the Earth. It explains the craters, the dust, and the difference between the two faces of the Moon.

THE INFLUENCE OF THE MOON

Since its formation, the Moon has had extraordinary effects on the Earth, all flowing from the influence of the Moon's gravity. It has lengthened the day, changed the spin, and created the tides, which were a crucial element in the evolution of life.

At first, before the coming of the oceans, the Moon caused tides in the rock – molten rock on the crust, molten rock in the core and, after the outer rocks hardened, slight distortions in the crust. This introduced a drag on the Earth's spin, slowing it down. Back then, some 4.6 billion years ago, the Earth was spinning over four times faster than now. The day was a mere 5 hours long. The effect of the Moon's drag is to slow the Earth's spin so any point on the Earth's surface receives light from the Sun for longer. This effect stretches the day by almost $1^1/_2$ seconds every 100 000 years:

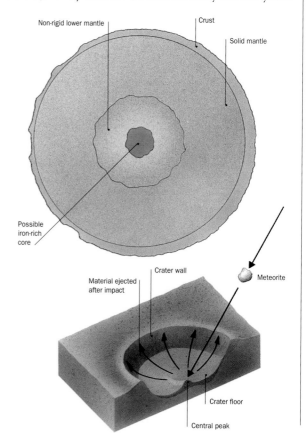

Non-rigid lower mantle

Crust

Solid mantle

Possible iron-rich core

Material ejected after impact

Crater wall

Meteorite

Crater floor

Central peak

THE LUNAR LANDSCAPE

The Moon was a desolate, dead world long before the first living cells appeared on Earth some 3.5 billion years ago. The lack of an atmosphere – any gas soon escaped the Moon's low gravitational field – and the absence of much geological activity have preserved surface features that on Earth would have been buckled and eroded out of existence.

The largest lunar features are the maria, or seas – flat, dark plains of basaltic rock – such as the Mare Serenitatis, and the Mare Tranquilitatis where Apollo 11 landed. These were blasted out by massive meteorites some 4.6 billion years ago. The molten rock that filled the resulting craters was very fluid, filling every indentation, creek and gulf in the crater walls. This created clearly defined 'coastlines' or boundaries, leading medieval astronomers to think the craters were seas. The larger craters, which are up to 150 miles (250 km) wide, have central peaks and walled sides.

At the same time, the impacts threw up great mountain ranges, such as the Apennines that border the Mare Imbrium, and spread liquid rock into strata that can still be seen.

Within 1 billion years, the Moon had cooled and the larger chunks of interplanetary debris had been blotted up. Lava flowed no more (though signs of its channels endure) and large impacts became rare. But smaller rocks – the size and frequency of impact decreasing with time – have since gouged the pitted surface of craters seen today.

A steady rain of micrometeorites has imposed a form of erosion unknown on Earth, scouring 6 ft (2 m) of rock off high points and depositing it in the lowlands, where it forms the dusty surface on which the American astronauts left their footprints.

MAKING AN IMPACT *Apollo 17 astronaut Jack Schmitt examines a huge boulder thrown up by a meteorite impact. He also discovered soil burned orange by the impact.*

that is 14 seconds every million years; 19 hours in the course of the Earth's history to date. As the Earth's spin continues to slow down, at some point in the far distant future the day will more than double in length.

By then, the night sky will be a very different place, as it was once in the past, because the tidal drag on the Earth has a further effect. The decrease in the Earth's speed is matched by an *increase* in the Moon's speed. This means that the Moon has always been moving away from the Earth, at the rate of 1½ in (4 cm) a year. Once, about 1 billion years ago, it lay a mere 11 000 miles (17 700 km) away, whipping around the Earth to give a month that was only 6½ hours long. Even when the first land animals emerged, it was orbiting at half its current distance, looming up at ten times its current size. Because it appeared larger from Earth, it would have blocked out the Sun for longer during eclipses. Total solar eclipses then would have been far more ominous, lasting a good two hours and casting a gloom over half the Earth.

There will come a time, some 1 billion years hence, when the day will have

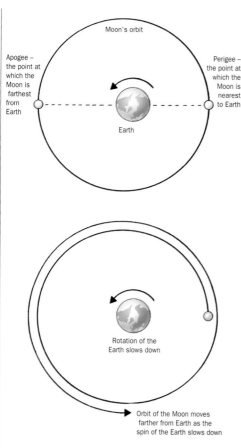

MOVING APART *As the Earth's spin slows down, the Moon moves away from the Earth at a rate of 13 ft (4 m) per century.*

stretched so much due to the slowing of the Earth's spin, and the Moon will be so far away and therefore take such a long time to circle the Earth, that a day and a month will be equal in length: both will be 55 of to-day's days long. The Moon by then will look like a tiny disc with no details visible, like a coin spinning across a garden – if there is anyone around to see it. Solar eclipses will be things of the past. The Earth will be a strange world, with tides (if there still are tides) a mere fraction of today's, inching in and out over a time scale 200 times longer, and a rising Sun that takes not two minutes to clear the horizon, but almost two hours.

Another consequence of the Moon's presence arose from the fact that the Earth was set spinning at a kilter, just over 23° off true. Its spin – which was very much faster

then than now – produced an equatorial bulge. The bulge was not on the same plane as the Moon's orbit, and so did not coincide with the Moon's pull. In fact, ever since the two took up their positions, they have been circling each other in a state of tension, with the Moon constantly tugging at the Earth's thickened waistline, as if it were trying to drag the Earth upright. The Earth, being so much more massive, resist-ed – and began to wobble. The effect is fa-miliar to anyone who has played with a gyroscope, or even a child's spinning top. A nudge causes the axis to wander, in a way or ways separate from its initial spin.

ECCENTRICITIES IN THE EARTH'S SPIN

Now we can return to the second and third of the spinning Earth's eccentricities, as analysed by the Serbian engineer, Milutin Milankovitch (the first being the 100 000 year cycle during which the Earth's orbit stretches from a near-circle to an ellipse and back, due to the gravitational pull of the other planets). The second eccentricity

involves a back-and-forth nod in the Earth's axis, a sequence that takes about 40 000 years to complete. The third eccentricity is one in which the axis moves in a slow circle, which takes just less than 26 000 years to complete. This motion, in particular, has in-teresting implications for anyone studying the so-called fixed stars.

The Sun, Moon and planets all seem to move along a narrow 8° band which runs through the 12 star patterns that make up the signs of the zodiac. As the Earth moves round the Sun, so the stellar background changes, which is why at roughly monthly intervals the Sun, Moon and planets are 'in' – that is, seen against – the different zodia-cal constellations. The zodiacal signs repre-sent an early attempt to understand the

GRAVITATIONAL PULL *The pull of the Sun and Moon deform the oceans by creating a tidal bulge on the side of the Earth facing them. Centrifugal force creates another tidal bulge on the opposite side.*

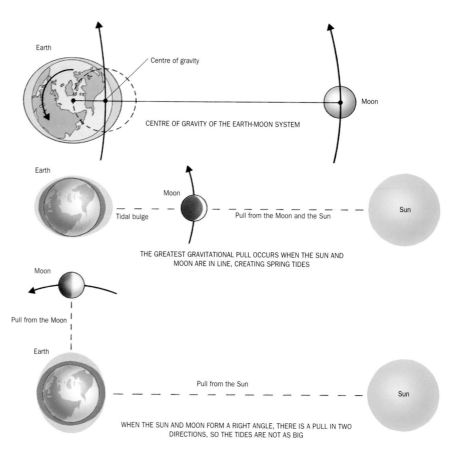

POLAR STAR TRAILS *A timed-exposure of polar stars reveals how the rotating Earth points towards the Pole Star, Polaris.*

provided by the constellation of Aquarius. In practice, of course, the Sun is so bright that the constellation itself, like any stars, can only be seen before sunrise and after sunset.

However, since the background of 'fixed' stars is slowly moving eastwards, the Sun's equinoctial position wanders westwards, or 'precesses' – hence the name given to this eccentricity in the Earth's orbit, the 'precession of the equinoxes'. Some 5000 years ago the Sun was 'in Taurus'; 2000 years ago it was 'in Aries'; and right now it has just moved into Aquarius from Pisces.

Precession introduces hideous complexities into astronomical calculations. (Indeed, experts have to take into account yet another eccentricity imposed on precession by the $18^1/_2$ year cycle in the Moon's orbit, which adds a slight nod, or 'nutation', to the Earth's spin.) All this can be ignored in the time scale of a human life, but is of great significance historically. The Mesopotamians, the Egyptians and the Greeks (who discovered precession) all made detailed observations. And the first two based religions, systems of government and massive architectural projects – most notably the pyramids in Egypt – on these observations.

This explanation of precession brings us back to the cycles analysed by Milankovitch: the 100 000 year cycle imposed by the stretching of the Earth's orbit; the 40 000 year nodding motion; and the 26 000 year cycle imposed by the precession of the equinoxes. Together, the last two minutely, but crucially, affect the amount of heat that falls on the Earth from the Sun, with effects that are part and parcel of the Earth's unique character.

movements of the heavenly bodies in terms of regular motions. The trouble is that, because of the Moon's influence, these movements are not regular. The Greeks discovered that the whole zodiac (indeed the whole firmament of background stars) changes minutely – by just less than $^1/_{100}$ of a degree – from year to year. Polaris, known as the Pole Star, once was not; and will not be again in the future – in AD 14000 the North Pole will be near the bright star, Vega.

This effect is commonly measured by seeing where the Sun rises and sets with respect to the background stars on a particularly significant day, the equinox, the day that occurs

PRECESSION *The gravitational pull of the Sun and Moon causes the Earth's axis of rotation to trace a circle.*

every six months when day and night are of equal length. Drawing on an astrological tradition, astronomers in the Northern Hemisphere regard the spring, or vernal, equinox (March 21 – the autumn equinox in the Southern Hemisphere) as the fixed point. At present the Sun is 'in Aquarius', which means that at the spring equinox it is positioned against the background

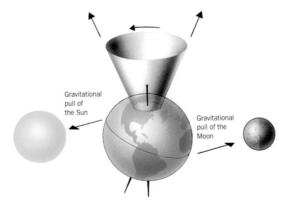

Gravitational pull of the Sun

Gravitational pull of the Moon

LAYING FOUNDATIONS FOR LIFE

Almost from the time it formed, the Earth's hot core and slowly seething outer mantle made the planet a crucible for the transformation of its basic constituents, creating ever more complex materials, from minerals to living cells.

In our biography of the Earth, we can imagine it a few hundred million years after its formation, a ball of rock the same size as it is today, but spinning much faster, continuously stressed by the looming presence of the newly formed Moon. Its day was shorter, its months shorter, but its year was the same length, because its orbit round the Sun was as it is now. Its first atmosphere, a volatile mix of hydrogen and helium forced out of the molten rock, had long since vanished into space. The core, already formed, was a seething mass heating the mantle and driving volcanic plumes up through a crust that was hardening but still too malleable and too distorted by the Moon's tidal pull to settle.

Already, a core mechanism was in place that would act as the Earth's heart. In fact, the Earth's core would have been hotter than it is now, because the radioactive material that drives it was only recently formed and had not had a chance to decay. So beneath the crust, the mantle incorporated treacly convection currents driving up from the centre, carrying heat towards the surface, cooling, turning and disappearing again into the depths. This rising heat churned the forming crust, producing hellish scenes – erupting volcanoes, splitting rock, violent earthquakes. Already, the lighter rock would have been cast up from the mantle, forming plates that rode like rafts of slag on top of the heavier rock beneath. These would one day become continents. They were separated by plains of basaltic lava, the future oceans, which grew from central splits in the crust and shrank as the plates of granite were carried back and forth on the convection currents below.

DRIFTING CONTINENTS

How do we know all this about events so long ago that the Earth's surface holds no trace of what happened? Because the mechanism established then – the means by which the surface was remoulded – is still operating now. The discovery of how it works was one of the most important scientific insights of the 20th century. In hindsight,

LAVA FLOW *In one of the most ancient geological processes, molten lava flows out of a volcanic vent, then cools and builds rock on the surface.*

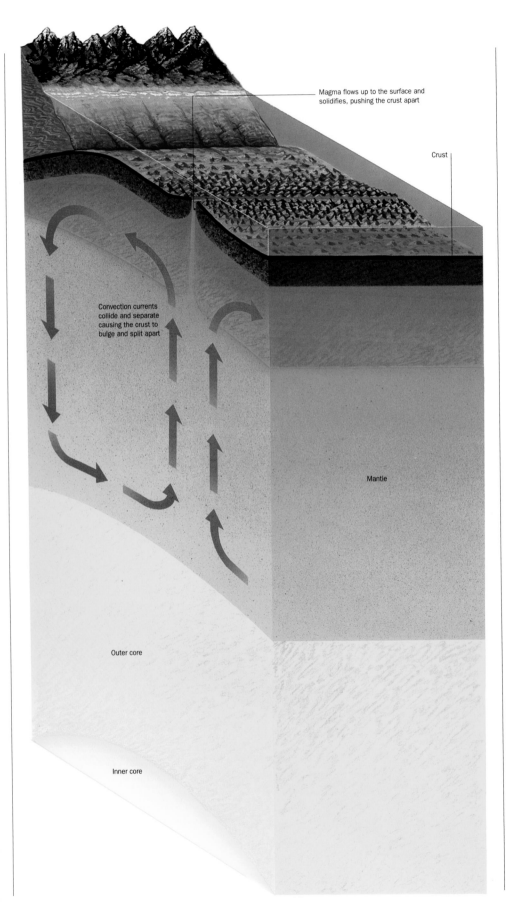

Magma flows up to the surface and
solidifies, pushing the crust apart

Crust

Convection currents
collide and separate
causing the crust to
bulge and split apart

Mantle

Outer core

Inner core

MOVING APART *At weak points
in the Earth's crust, such as at
mid-ocean ridges, convection
currents circulating in the
Earth's mantle cause the crust
to split and move apart.*

of course, it seems obvious. A glance at any
atlas shows the apparent fit between the
bulge of South America and the bight of
Africa. Once, the idea of massive Earth movements was the stuff of legends like that of
Atlantis. But as modern science developed,
myth was dismissed, and the idea of continents moving became heresy. Geology was
dominated by the idea that land could rise
and fall, but not move sideways.

The man who established continental
drift as a serious scientific proposition was
the German meteorologist Alfred Wegener:
he did so in two articles published in 1912.
He concluded that until 40 million years
ago all the continents were joined into one
landmass, which he called Pangaea ('Land
Everywhere'). Pangaea was like a grounded
ice floe, which then broke up, allowing the
separate continental blocks to float away on
a partially liquefied layer beneath.

Few people took Wegener seriously, and
with good reason because he had no idea
of the mantle's convection currents and
could not propose a mechanism that would
actually move the plates. Only in the 1960s
did Wegener's ideas receive proper attention, initiating a revolution as profound as
the one triggered in the second half of the
19th century by Charles Darwin's theory of
the evolution of species. Marine geologists
and geophysicists, such as Maurice Ewing
and his co-workers at Columbia University,
discovered that the crust under the oceans
was only a few miles deep, much less than
the crust under the continents. They also
discovered that a mid-ocean ridge, first
seen in the Atlantic, was a feature of all
major ocean basins.

In an attempt to explain these observations, Professor Harry Hess of Princeton
University in 1960 proposed the idea that
heat flowed up from deep in the Earth,
creating convection currents in the
mantle's rock, which emerged in crests in

the middle of the oceans. As the currents hit the surface, they spread away from the ridges, carrying the ocean crust with them. The gap was filled by rock welling up from below. Hess suggested that along each ridge about 1/2 in (1 cm) of new crust is formed each year. At this rate it would take a mere 200 million years for all the present deep-ocean floor to form. Since this is only about 5 per cent of the Earth's age, he also suggested that the old crust is destroyed at the same rate as the new crust is generated.

Confirmation of Hess's ideas came from observations of the magnetisation of the sea-floor rocks, which record the direction of the Earth's magnetic field. Since that field is not fixed, but flips direction now

and then, the new crust builds up on either side of the ridge in parallel stripes that are alternately magnetised, in line with and opposite to the Earth's present magnetic field. Clearly, if the sea floor is spreading, the continents along its boundaries must be moving. Quite suddenly, in the early 1960s, the idea of continental drift, which had been so roundly rejected decades before, became the new orthodoxy.

What now emerged was the so-called theory of plate tectonics, which differs from earlier propositions in seeing the moving units as involving much more than the continental crust. As the Earth settled towards its present form, the top section of the mantle and the crust formed a rigid unit known as the lithosphere, while the lower

section of the mantle – the asthenosphere – remained a very slow-moving liquid. The lithosphere broke into seven great slabs, or plates, and several lesser ones. The plates were of two kinds, thick continental plates and thinner plates on which the oceans lay. The plates were moved about by the asthenosphere's convection currents.

As the plates moved around, they interacted with one another along their boundaries in different ways. One way was for two plates to move apart at a spreading ocean ridge, where new crust material was formed. Another happened in areas where two continental plates collided: both plates crumpled and threw up huge mountain ranges. In some places, two plates did not collide, but slid past each

PUSH AND PULL
Plate boundaries occur either where two plates converge, colliding with or sliding past each other, known as a convergent plate boundary; or where a plate splits and spreads apart, known as a divergent plate boundary.

Continental plates are too light to subside into the mantle. Where two converge, they are compressed together, pushing up mountain ranges along their edges. Eventually, the two plates fuse into one.

Another form of convergent plate boundary occurs where a continental plate meets an oceanic plate. The continental plate, which is the lighter of the two, slides over the top of the heavier oceanic plate, the edge of which is pushed down into the Earth's mantle.

Where a plate splits, a fissure opens up along the fault, and the two sections move apart, forming a divergent plate boundary. Magma wells up through the split, then cools and becomes part of the plate. Most divergent boundaries occur on the ocean floor.

WEGENER AND THE DISCOVERY OF 'CONTINENTAL DRIFT'

Today, continental drift seems an obvious explanation for the lie of the world's landmasses. Yet the facts, and the theory behind them, were accepted only in the 1960s, long after the idea was first proposed by Alfred Wegener.

Wegener, a German, was not a geologist at all but a meteorologist. 'The first concept of continental drift came to me,' he wrote, 'as far back as 1910, when considering a map of the world, under the direct impression produced by the congruence of the coastlines on either side of the Atlantic.' Doubtful at first of his own idea, he began to gather evidence that land now divided by ocean had once been joined.

The evidence drew on different disciplines. During the Permian Age, for instance, the Southern Hemisphere had been covered by ice. If the continents were in their present position, the extent of the ice must have been so vast that it would have taken all the water on Earth to make

ON THE MOVE *In* The Origin of the Continents and Oceans *Wegener published a series of maps to illustrate the way in which he thought the continents once meshed together.*

PIONEERING THEORIST *Alfred Wegener, the first scientist to propose the idea of continental drift. He died before his theory became accepted.*

it. At the same time, the Northern Hemisphere seemed to have been tropical, covered by rain forests that later formed rich coalfields. Did it make sense that the world was cold on one side and hot on the other?

Finally, between ice ages in the south, a particular assortment of plants grew, dominated by the giant fern *Glossopteris*, now preserved in low-grade coal in South America, Antarctica, Africa, Australia and India. The same plants, and the same combinations, appeared in all these places, a sequence named after Gondwana in India.

Wegener was also impressed by the evidence of fossil animals. A species of garden snail lived in both North America and western Europe; identical worms were found on both sides of the Atlantic, related marsupials in Australia and South America.

All this could only make sense – as Wegener suggested in two articles in 1912 and then in *The Origin of the Continents and Oceans* (1915) – if there had been one supercontinent, Gondwanaland, that broke up some 200 million years ago. As a theory it explained much, but lacked an explanation itself. Wegener imagined the continents barging their way through the ocean floor like ships in pack ice, with a bow-wave of earthquakes and mountains. But what moved the continents?

Experts found it easy to dismiss

him. The idea was 'contrary to all available physical and geological evidence', wrote one geologist, Walter Bucher. The crystallographer Sir Lawrence Bragg claimed that the time he showed some of Wegener's work to a well-known geologist was the only occasion when he actually saw someone foam at the mouth.

It was only in the early 1960s that continental drift acquired its theoretical foundations in the theory of plate tectonics, with a growing understanding of slowly seething columns of molten rock that emerge in oceanic ridges and carry the lighter-weight continental plates around the surface of the Earth.

Wegener, the focus of such furious controversy, never lived to see his ideas vindicated. He died while exploring Greenland in 1930.

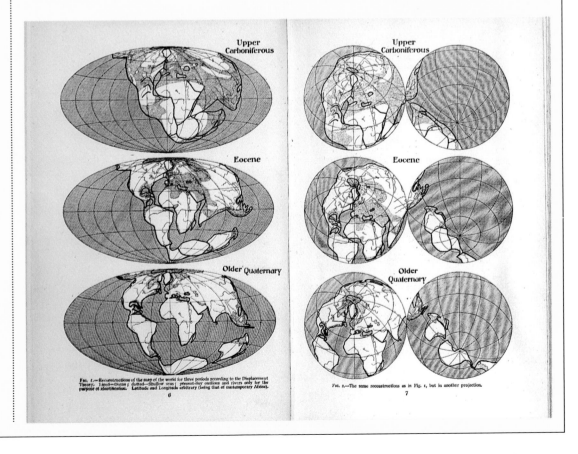

other, without plate material being either created or destroyed.

Finally, there was the type of boundary where oceanic and continental plates collided. The continental plate would be thicker than the oceanic plate, which would be overridden, buckled and forced downwards in a so-called 'subduction zone', leaving a trench along the ocean floor. Some 400-500 miles (650-800 km) down, the oceanic plate would break up. But being lighter than the surrounding mantle, it would form a slag – lava – that would rise back to the surface, emerging through mid-ocean ridges and

INVISIBLE SEAM *This small ravine in Iceland is the seam where the American continental plate and the European continental plate meet, and along which they are being torn apart.*

volcanoes. Along the boundary where the two plates met would be a long zone of intense earthquake activity.

In this model, the continental plates became the truly enduring units of the Earth's changing surface. The oceanic plates, the ones that were consumed in subduction zones, would eventually have been totally eaten up, leaving two continental plates to collide and replace them.

Since the continental plates emerged some 4 billion years ago, they have been jostled back and forth across the surface many times, and the present arrangement offers few clues to what happened in the remote past. The Appalachians, the ancient, eroded stumps of once-great mountains stretching south-west from the St Lawrence River in Canada to the US state of Alabama, are the remnants of a collision between America and Europe some 360 million years ago, when the gap previously occupied by some primitive Atlantic was destroyed. But that takes us only a quarter of the way back towards the Earth's youth. Ancient rocks survive from those days, in Australia, Greenland, northern Canada and southern Africa, but so eroded, buckled and displaced that they offer no guidance

ANCIENT SITE *Under Greenland's snow and ice lie rocks that are ancient survivors from the Earth's youth.*

on the look of the Earth with its freshly minted continents and oceans.

The Earth's early, tortured state, both before and after the crust began to harden, would have been intensified by the fearful impacts of meteorites left over from the foundation of the Earth-Moon system and of the Solar System itself. The bombardment would have matched the Moon's. Asteroids many miles across rained down, shattering the forming crust, creating craters that were dozens, or hundreds, of miles across. For instance, an asteroid 6 miles (10 km) wide would have blasted a crater 200 miles (320 km) across and splashed rock from the Earth out into space.

All signs of this assault vanished quickly as the mantle and crust re-formed. But the evidence for it is there on the Moon. The first 'seas' on Earth must have been like the lunar ones – gaping, lava-filled craters. As time passed, they were overlaid and blurred by later impacts, then spattered with the

decreasing rain of debris left over from the process of creation, until the drifting continental plates rubbed them out completely.

All the inner planets were subject to such massive impacts. As a result there would have been a steady scattering of the planetary rocks back into the inner Solar System. Some of these rocks would eventually be caught by the gravitational fields of the Moon or other planets. The Earth did not evolve in isolation: somewhere on Earth, there are bits of Mercury, Venus, Mars and the Moon. Somewhere on all these bodies there are chunks of Earth.

OCEANS AND AIR

The chief products from the volcanic instability and the ceaseless bombardment from space were carbon dioxide and water vapour, as well as smaller quantities of hydrogen, sulphur dioxide, and possibly methane and ammonia. It would not have taken long – perhaps as little as 100 million years, $1/50$th of the planet's history – for this, the Earth's second atmosphere, to form.

For the next few hundred million years, while the Earth's

ERODED STUMPS
The Appalachian Mountains are the remnants of peaks thrust up by a continental collision some 360 million years ago.

4.5 BILLION YEARS AGO   VOLCANIC ORIGIN OF EARTH'S FIRST ATMOSPHERE

Water, hydrogen, carbon dioxide and other gas

Earth too hot for water to condense

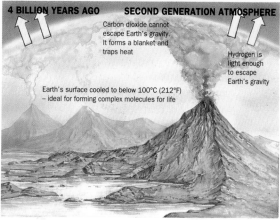

4 BILLION YEARS AGO   SECOND GENERATION ATMOSPHERE

Carbon dioxide cannot escape Earth's gravity. It forms a blanket and traps heat

Hydrogen is light enough to escape Earth's gravity

Earth's surface cooled to below 100°C (212°F) – ideal for forming complex molecules for life

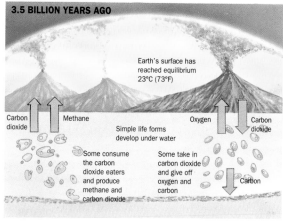

3.5 BILLION YEARS AGO

Earth's surface has reached equilibrium 23°C (73°F)

Carbon dioxide   Methane   Oxygen   Carbon dioxide

Simple life forms develop under water

Some consume the carbon dioxide eaters and produce methane and carbon dioxide

Some take in carbon dioxide and give off oxygen and carbon

Carbon

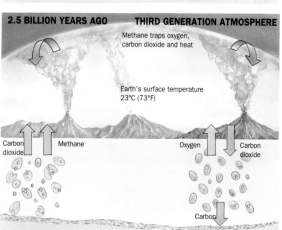

2.5 BILLION YEARS AGO   THIRD GENERATION ATMOSPHERE

Methane traps oxygen, carbon dioxide and heat

Earth's surface temperature 23°C (73°F)

Carbon dioxide   Methane   Oxygen   Carbon dioxide

Carbon

surface remained above the boiling point of water, the water vapour would have had no chance to condense. But as soon as the crust cooled to below 100°C (212°F) – perhaps some 4 billion years ago – water would have begun to form and fill the low-lying areas that would one day hold the oceans. For millions of years, the process continued. If the past was roughly like the present, volcanoes would have 'outgassed' some 10 million tons of water vapour every year, enough to account for all the water in the oceans today. At last, some 3.5 billion years ago, steady cooling led to equilibrium.

At that point, the Earth would have been hot, but probably not unbearably so, according to a 1979 study published in the USA in the magazine *Nature* by three experts in atmospheric chemistry and climatology, T. Owen, R.D. Cess and V. Ramanathan. They assumed that, since stars grow hotter as they age, the Sun's heat output was 25 per cent less than it is now. They estimated the amount of carbon dioxide that would have been ejected from the Earth's interior to be at least 200 times, and perhaps 1000 times, as much as there is in today's atmosphere. Carbon dioxide is a greenhouse gas – it lets the Sun's rays through, and traps

THE CHANGING ATMOSPHERE
Both volcanic activity and early life forms contributed to the mix of gases in Earth's atmosphere, changing it from carbon-dioxide rich to methane rich, and then to oxygen rich, paving the way for the evolution of complex oxygen-consuming life forms.

the heat at the Earth's surface by absorbing infrared radiation. This would have compensated for the Sun's lower output. They concluded that the average temperature was 23°C (73°F) – about the same as the tropics today.

This would not have been uniform, of course, because the polar regions would receive far less heat. Here, already, water vapour would have created ice. From afar, the Earth would have revealed a hint of its chemical and geological variety, with the white of the poles overlaid on the greys and browns of its bare rocks.

All these conditions were vital for the Earth's future. In particular, it now had an atmosphere and water, both of which would offer protection against some of the Sun's most damaging radiation. The water offered a solvent within which molecules could meet, mix and react. The temperature range was mostly within the limits that govern the formation of more complex molecules (for freezing locks up chemicals, while boiling disperses them).

Crucially, as far as future life was concerned, there would have been no oxygen in the atmosphere. The oceans would have been rich in iron eroded from the bare land. The substances and gases present in the surface of the young Earth readily combined with oxygen, so any that emerged was soon blotted up. In turn, this meant that there was no layer of ozone, the oxygen derivative that now protects all life from ultraviolet radiation.

THE COMING OF LIFE
Yet life arose, and did so remarkably quickly. It has become almost commonplace for anyone discussing the evolution of the Earth and its life forms to dramatise the time scale by compressing it into a single day. On this scale, the dinosaurs emerge after 23 hours, 58 minutes and 48 seconds. All of human life is compressed into the last 3.6 seconds. The impression given by this image is that life is a rather belated occurrence, but this is not so. Life on Earth started almost immediately it became possible. Two sets of rocks have been discovered in Western Australia, one with a dark,

EARLY INHABITANTS *The algae and bacteria that colour this Wyoming geyser are descendants of the very first life forms that appeared some 3.5 billion years ago.*

Since then, scientists have found that amino acids are formed from constituents of the Earth's early atmosphere under a wide range of stimuli. Electric shocks are the most effective, as presumably lightning must have been – and lightning storms may have been a far greater feature of the young Earth, with its volcanic landscape and aeons of rain, than they are now. But other stimuli also work, such as shock waves and ultraviolet light, artificial equivalents of thunderclaps and solar radiation.

Of course, such jolts destroy as well as create, and the Earth's solar radiation would have broken apart countless products from the planetary test tube. Without an ozone layer, the intensity of ultraviolet light would have been 30 times higher than it is now.

Once it was argued that the ozone layer was essential to all life, but this is probably

layered structure that could have been made by bacteria, the other containing microscopic, shadowy, worm-like shapes that resemble modern bacteria and round cells like a type of modern algae. Both rocks date from about 3.5 billion years ago.

Later on, life was to play an intimate role in the evolution of the Earth, in effect creating its own environment. At the start, though, it simply arose and stayed. How it arose, scientists have yet to explain in detail, but there are some well-founded theories.

All known living systems are composed of carbon and water, and need energy, in the form of sunlight, to sustain them. The basic chemical constituents build successively into amino acids, proteins and nucleic acids, which encode the genetic instructions of cells. This is a long way from life itself, but theory and experiment combined as long ago as the 1930s to suggest that there is a continuity between non-life and life.

In 1953 Stanley Miller, an American biochemist, conducted a classic experiment at the University of Chicago on the possible origin of life on Earth. He assembled a mixture of hydrogen, methane, ammonia and water vapour, some of the gases that are believed to have formed the Earth's second atmosphere (after the loss to space of the first one composed of hydrogen and helium). He theorised that with an available power-source – lightning – the primitive atmosphere could have been made to synthesise more complex chemicals. To test this idea, he passed an electric spark through his artificial atmosphere. At the end of a week, he had produced a wide range of complex organic molecules, including some amino acids – molecules that are vital to life because they form chains that make up proteins, the building blocks of all living things.

BUILDING BLOCK OF LIFE
An 80x magnification shows crystals of glycine, the simplest and probably the earliest amino acid.

A NEW WORLD FROM BENEATH THE WAVES

The mechanism that causes continents to drift apart – the steady emergence of new rock along a split in the ocean floor – can sometimes appear dramatically on the surface. On such occasions, the eruption also exemplifies the way volcanoes contribute to atmospheric change, as well as providing opportunities for new life.

On November 14, 1963, an Icelandic fishing boat near the Westman Islands, 10 miles (16 km) from Iceland's south-west coast, reported an explosion. The following night, beneath a column of smoke 6 miles (10 km) high, and accompanied by further explosions and a rain of pumice, a cauldron of black lava emerged from the depths. The Westman Islands were in the process of acquiring a new member, Surtsey, which became a visible part of the Mid-Atlantic Ridge, the line that marks the geological division between North America and Europe.

By April the following year, when the eruption was over, the island was large enough to be included on maps as a permanent feature. That summer scientists visiting the new arrival noticed that its surface was no longer entirely barren. Seeds had arrived from the mainland. Surtsey, with its first fern and its first butterfly, was already alive.

NEW LAND, NEW LIFE *Within a year of appearing, Surtsey's surface of volcanic ash was already host to plant life.*

FAMILY TRADITION *Like their predecessors, these algae in the Waimangu thermal springs, New Zealand, live in warm water, absorbing hydrogen and releasing oxygen.*

not so. The biologist James Lovelock describes an experiment in which he and other scientists, in an attempt to create a sterile environment in hospitals, experimented with ultraviolet radiation as a way of killing bacteria. It worked – but only if the bacteria were suspended in air, entirely unprotected. 'In the real world,' Lovelock writes, 'they do not live naked any more than we do.' Their own secretions, or a thin film of mineral-rich water, were enough to shield them.

Many scientists now accept that with so many ways in which organic molecules can be made, and with a planet-wide blanket of water to act as a protection, the seas of the primitive Earth became a sort of chemical soup from which life could have arisen.

LIFE FROM SPACE?

There is one other possibility – one that sounds more like science fiction than science fact, but which is hard to discount. Perhaps life arrived on Earth from space.

The ingredients are certainly there. It has been known since the 1930s that atoms in space combine to form simple molecules. Then, in the 1960s, when radio astronomy was developed, astronomers found a whole range of more complex molecules, many incorporating carbon, which makes them possible constituents of organic compounds. In addition, it can

hardly be a coincidence that the most common molecules in the Universe – carbon, hydrogen, nitrogen and oxygen – are also the most abundant in living creatures on Earth. To date, more than 80 types of molecule have been found in space, suggesting that the building blocks of life (though not life itself) may be a feature of the Universe, not simply of Earth.

What of more complex molecules still – like amino acids, the building blocks of proteins? These, too, have been found. On the morning of September 28, 1969, a meteorite fell at Murchison, 100 miles (160 km) north of Melbourne in Australia, and was identified by John Lovering, then Professor of Geology at Melbourne University, as a type of meteorite with little 'chondrules' of silicate material embedded in it. Such meteorites are named carbonaceous chondrites, from their carbon-rich content. The Murchison meteorite contained five amino acids common to living protein.

The notion that life might have developed in deep space was formulated by the eminent astronomer Sir Fred Hoyle and Professor Chandra Wickramasinghe of Uni-

LIFE IN SPACE *The surface of the meteorite that fell in Murchison, Australia, in 1969, contained fibres formed by organic materials in deep space.*

versity College, Cardiff, in the 1980s. The two had spearheaded research into interstellar dust, and knew from spectral analysis that organic molecules existed in space. They could not connect the two observations, however, 'until one day in 1979 when we arrived logically at the startling conclusion that cosmic dust had to be freeze-dried bacteria'. According to this view, life is a property of the Universe, emerging as a natural consequence of the endless cycle of stellar creation and destruction, scattered across the galaxies by drifting clouds, the pressure of light, and meandering bodies such as comets. Hoyle and Wickramasinghe proposed that life was seeded onto Earth by comets.

Although this remains a highly controversial theory, there is no shortage of the right raw materials, either on Earth or

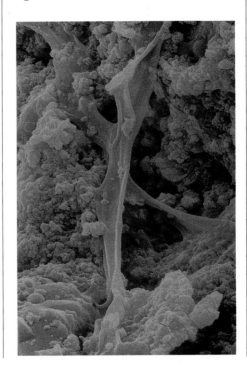

throughout the Galaxy. It also has the great advantage that the test tube from which life emerged – the creation of vastly unlikely and complex molecules – is no longer a test tube the size of the Earth, but one that is billions of times larger: the Galaxy as a whole. Finally, there is evidence that micro-organisms can survive in space: in the 1970s, a living bacillus was recovered from a TV camera that had been left on the Moon for two years, and others have endured pressures of up to 65 tons per sq in (10 tons per cm²) and temperatures of up to 600°C (1112°F).

EARTH'S GROWING COMPLEXITY

Wherever it came from, life established a hold, with early organisms that were a type of bacteria using carbon dioxide as food, fermenting to break the gas down into carbon and oxygen. A possible sequence of events has been proposed by James Lovelock, who argues that the very young Earth evolved a self-regulating complexity that led on from an oxygen-poor world, hostile to life as we know it, to the oxygen-rich world in which later life forms thrived.

Lovelock argues that the bacteria that consumed carbon dioxide could not have been the only organisms, because otherwise their numbers would have exploded and they would have gobbled up all the carbon dioxide, removing Earth's protective blanket and initiating an ice age from which there would have been no escape. The carbon dioxide they consumed must have been returned to the environment in some way, probably by another form of bacteria, which would live on the carbon-dioxide-consuming bacteria and convert their remains back into carbon dioxide, secreting methane in the process.

Methane itself is a greenhouse gas. It would have risen high in the atmosphere, replacing the carbon dioxide and also trapping the oxygen that was produced from the Earth's interior. That changeover – to the third generation of terrestrial atmosphere – could have occurred quite suddenly around 2.5 billion years ago.

At this point, a visiting alien might have looked down on a brownish-red planet and

VERSATILE SURVIVORS
Bacteria, which helped modify the Earth's early atmosphere, survive in marshes, where their methane products sometimes flare as will o' the wisps (right). Others thrive in the digestive tracts of grazers such as wildebeests (below).

jumped to the conclusion that it was as dead as Mars. But an air sample would have revealed that something dramatic was happening. No chemical reaction would explain the rise in methane and oxygen, or the overall balance between the gases. The alien would have known that the Earth was now alive.

A landing would confirm it. Down on a seashore, the Sun would be less bright than now, with an orange glow, like a present-day sunset. The sky would have a pink tinge. No tracks would mark the sand, of course, but with a closer look, promising signs would appear. The low tide would reveal odd, mushroom-shaped, coral-like structures formed by the calcium carbonate secreted by bacteria. Inland ponds would have green and black patches. There would be a bad-egg smell of methane, and the plop of methane bubbles bursting from the marshy

edges of the ponds. Across the higher ground, perhaps, would be a faint varnish of microbial life, rooted in rock and living off carbon dioxide, continually absorbing, changing and secreting the nutrients present in the rock.

The coming of oxygen around 2.5 billion years ago marked the beginning of the Earth as we know it today, but it did not mark an end for those organisms that evolved in an oxygen-free world. They live on still, scavenging on the sea floor, in marshes (where their noxious methane secretions sometimes take fire as will o' the wisps), and in the guts of higher animals, where they perform their ancient role of breaking down carbon-rich food and releasing methane.

Meanwhile, all around, a new world had grown up, with a different chemistry and an explosion of possibilities.

A Machine Called Earth

4

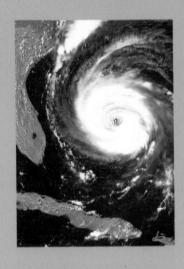

DEATH SPIRAL *Hurricane Fran approaches the USA in September 1996.*

THOUGH DISASTERS SUCH AS EARTHQUAKES, VOLCANIC ERUPTIONS, FLOODS AND DROUGHTS ROUTINELY SPREAD DEVASTATION AMONG HUMAN POPULATIONS, TO THE EARTH THEY ARE MERE BLIPS SET AGAINST AN UNDERLYING PATTERN OF STABILITY. THAT STABILITY DEPENDS ON MECHANISMS THAT HAVE EVOLVED OVER AEONS. DURING THE PAST 3 BILLION YEARS, THERE HAVE BEEN MANY REVOLUTIONS THAT HAVE MODIFIED THE EARTH'S CHEMICAL AND GEOLOGICAL MACHINERY. EACH NEW STAGE IN THE DEVELOPMENT OF LANDMASSES, OCEANS AND ATMOSPHERE HAS OPENED UP NEW OPPORTUNITIES, ALL ENRICHING THE PLANET AND INCREASING THE POSSIBILITIES FOR THE GROWING NUMBERS OF EVOLVING LIFE FORMS.

NEW ROCK *Lava pours from Mount Kilauea, Hawaii.*

THE ONE AND ONLY PLANET

Behind the random extremes of heat, cold, wind, drought and deluge are the processes that define the character of the planet, providing climate as well as weather, and balancing catastrophe with stability.

The French architect Charles le Corbusier once said that a house was a machine for living in. The remark has remained controversial because, although houses can be seen as machines in many ways, they contain elements that are utterly random and very unlike a machine.

The same is true of the 'machine' on which we live, the Earth, whose dominant traits, whether predictable or random, have been part of the changing surface since the oceans first formed at least 3.8 billion years ago. It is a surprisingly early date, because in theory there should not have been any oceans at all. At that time, the Sun was some 25 per cent cooler than it is today, as astronomers know from comparing its output to other Sun-like stars. Calculations suggest that all the water on Earth at that time should have been frozen, not starting to melt until about 2 billion years ago. In fact, scientists know that the oceans were in existence then, because the oldest rocks are sedimentary rocks, laid down under water. So why was the water on Earth not frozen?

The answer is that the Earth's interior was hotter because the radioactivity deep within it was at a higher level. This would not have affected the surface temperature

FROM ICE TO AIR *Seen from above the Earth reveals a range of watery conditions, from Greenland's ice cap, a remnant of the last Ice Age (opposite), to oceans and cloud systems such as the cumulonimbus (right) rearing 9 miles (15 km) into the sky, giving warning of a storm to come.*

directly, but it would have lent more power to the convection currents circulating underneath the crust. That, in turn, would have meant greater instability and more volcanoes. Hence a preponderance of carbon dioxide, a greenhouse gas, ensuring that the otherwise chilly Earth had a warming blanket of gas trapping heat at the planet's surface.

Whether alien astronomers observed the Earth then or now, they would see comparable features – swirling clouds, huge oceans, continents, slow-moving ocean currents – and know that this chemical cauldron was almost certainly brewing life. Immense global processes, such as ocean currents and weather systems, ensure the production, interaction and circulation of elements and minerals. Obviously, this is a rich, complex and ever-changing planet. The problems alien observers would have faced are the same as scientists face now: how to quantify and how to understand what is happening, and how to separate the predictable from the random.

One example of this problem can be found in relation to the weather, whose vagaries, especially in mid-latitudes, are notorious. Indeed, they provided scientists with the inspiration for a whole new branch of science, a story told by the science writer James Gleick in his book *Chaos*.

In the 1960s, the American meteorologist Edward Lorenz was experimenting with computer programs, hoping that with enough information he would be able to predict weather as precisely as was possible for the motions of the planets. 'The average person,' he said, 'seeing that we can predict tides pretty well for a few months ahead

would say, why can't we do the same thing with the atmosphere, it's just a different fluid system.'

One day, he reran a program in order to re-examine its results, but to save typing time he defined his numbers to three decimal places only. The difference seemed insignificant – after all, thermometers did not measure the temperature of the ocean or air to an accuracy of $1/1000$ of a degree. However, to his astonishment, the results of the rerun program bore no relation to the original.

In a sudden flash of insight, he realised that long-range weather-forecasting according to a computer model was impossible – and always would be – because in complex systems such as the weather, minute irregularities multiply rapidly. No matter how close together the sensors picking up information are positioned, and no matter how quickly

they pick up information, there will always be minute changes, both near the sensors and during the intervening seconds, that go unrecorded. These changes cascade through the equations. Even one unknown could wreck a prediction, a realisation dramatised in a popular image: a butterfly flapping its wings in Peking transforms storm systems in New York a month later. It is this, the Butterfly Effect, that gives the Earth's weather its random character.

THE WEATHER MACHINE

As a consequence of the Butterfly Effect, is our impression that there are underlying rules to the weather an illusion, and should we discard the whole notion of a weather machine? Not at all, for there are underlying mechanisms. It is the way in which they manifest themselves that contains the randomness. This chapter looks

THE FLOOD: AN EYEWITNESS ACCOUNT

The story of the Flood that drowned the Sumerian civilisation of the Tigris-Euphrates valley has had a mixed history, being accepted as literal truth and dismissed as mere myth. Archaeological research has now revealed many different prehistoric floods. Perhaps one was so overwhelming that it entered folk history as the Flood, as portrayed not only in the Biblical story of Noah, but also in the Babylonian *Epic of Gilgamesh*, versions of which date back to about 2000 BC.

In this extract, the Sumerian Noah, Ut-Napishtim, has built the ark and awaits the onslaught. As a report of a hurricane driving a storm surge in from the Persian Gulf, it carries a dreadful authenticity.

'The time was fulfilled, the evening came, the rider of the storm sent down the rain. I looked out at the weather and it was terrible, so I too boarded the boat and battened her down. All was now complete, the battening and the caulking; so I handed the tiller to Puzur-Amurri the steersman, with the navigation and the care of the whole boat.

'With the first light of dawn a black cloud came from the horizon: it thundered within where Adad, lord of the storm, was riding. In front over hill and plain Shullat and Hanish, heralds of the storm, led on. Then the gods of the abyss rose up; Nergal pulled out the dams of the nether waters, Ninurta the warlord threw down the dykes, and the seven judges of hell, the Annunaki, raised their torches, lighting the land with their livid flame. A stupor of despair went up to heaven when the god of the storm turned daylight into darkness, when he smashed the land like a cup. One whole day the tempest raged, gathering fury as it went. It poured over the people like the tides of battle: a man could not see his brother nor the people be seen from heaven. . . .

'For six days and nights the winds blew, torrent and tempest and flood overwhelmed the world, tempest and flood raged together like warring hosts.

'When the seventh day dawned the storm from the south subsided, the sea grew calm, the flood was stilled. I looked out at the face of the world, and there was silence.

'All mankind had turned to clay.'

ASSYRIAN HERO *Gilgamesh, portrayed in an 8th-century BC statue, sought the survivor of the flood, Ut-Napishtim, to learn how to overcome death.*

at these random manifestations in human terms, because it is human experience that acts as the source of information about these processes – processes that have operated continuously since the appearance of water and solid land.

To understand this unique combination of machine-like predictability and chaos, it is useful to build up from first principles. Imagine the Earth as a smooth, spherical planet with no oceans, without any spin, circling the Sun with absolute stability. It would be heated unevenly, with more heat falling on the Equator than the poles. Hot air would rise at the Equator, drawing in colder air from north and south. Convection would carry the hot air towards the poles and cold air towards the Equator to ensure a balance. That would be the whole story of the Earth's weather: a steady wind blowing towards the Equator from the poles.

The Earth, of course, is anything but a neat sphere. Oceans and continents produce different kinds of heating in the atmosphere. Winds are deflected by mountain ranges. They pick up moisture from the oceans and drop it over land. The Earth is spinning, giving a sideways twist to the wind patterns. And the Earth lies at a slant, imposing yet another variable on the heating patterns.

All of these processes can be seen as the result of the effect of energy – in the form of heat – from the Sun and from the Earth's interior arriving at the Earth's surface. As the energy flows into air, earth or water, it builds up, then dissipates, in ways so complex and random that, like the weather, they may always defy final analysis.

It is the Earth's tilt that imposes the longest phases of regularity on the changing weather patterns. It explains why winter is colder than summer, and why the seasons follow each other. The difference has nothing to do with any change in the distance from

the Earth to the Sun: although this changes by some 3 million miles (4.8 million km) in the course of a year, it makes no significant difference to the amount of heat received.

The true cause of the rhythm of the seasons is the Earth's tilt – 23.5 per cent at present, although it changes minutely with precession (caused by the slow, circular path followed by the Earth's axis) – which in the course of a year means that first one pole, then the other, leans towards the Sun. As seen from the surface of the Earth, the Sun appears more directly overhead in summer; and its rays are less diffused by the atmosphere. Moreover, the days are longer. Indeed, because of the tilt, in polar regions the Sun lies either permanently low in the sky or is never seen at all, giving a summer 'day' and a winter 'night' several months long.

But again, there is a further complexity. The peaks of heat in the summer and the troughs of cold in the winter lag behind the position of the Sun because of the time it takes for the Earth to heat up and cool down. In the Northern Hemisphere, for

SEASONAL ALTERNATION
The Earth's tilt ensures the succession of seasons in temperate regions from summer (below left) to winter (below).

MIDNIGHT SUN *At midnight in midsummer, the Sun shines down on an Inuit village in the Canadian arctic, where it never sets during summer.*

instance, the Sun is highest in the sky on June 22, but at this point the Sun's heat is still being absorbed by the air, oceans and land. The summer does not build to a peak of heat until July and August. In the autumn the heat dissipates slowly, the Earth acting as a huge storage heater. The shortest day, December 22, is often far from being the coldest.

In addition, the oceans have always masked the effects of seasonal changes. The seas warm up more slowly, and lose heat more slowly, than land. Areas surrounded by water – such as Britain – therefore have less extreme variations than those with landlocked heartlands, such as North America or Eurasia.

WINDS

Whatever the look of the shifting landmasses, whatever mountain ranges have risen and disappeared again over the millennia, there have always been traces of the steady winds of convection that would dominate a smooth planet. These are most obvious over the oceans, away from the distorting effects of landmasses. They appear as the winds that proved so useful to explorers and merchantmen sailing round the world. Without the Earth's spin, the 'trade winds', as they came to be known in the 18th century, would blow straight from north and south. But the east-west spin deflects them so that they blow from the east – from the north-east in the Northern Hemisphere and from the south-east below the Equator.

Far from the Equator, around the poles, the sea-level winds are dominated by cold air spreading down from the frozen regions. Again, these icy winds have an east-west bias. In the middle latitudes, the east-west airflows to north and south are counteracted by strong winds from west to east high in the atmosphere, which create swirling weather systems below.

This is the realm of the 'jet stream', a high-altitude, high-velocity torrent of air that is continuously unstable, its west-east trend constantly masked by the chaotic effects it creates. Sometimes the trend is quite regular; sometimes it flows in zigzags that scatter the influence of the weather systems over a much broader belt. The causes of these disruptions in the jet stream are obscure – distant volcanic eruptions, changes in sea-surface temperatures – but the effects can be catastrophic in human terms. In America's Midwest, 1988 was a year of drought and 1993 the year of the

Great Flood, leading to over 1000 river-wall collapses and the flooding of 13 million acres along the Mississippi, Missouri and their tributaries.

CLIMATE

The Earth's underlying stabilities – the Sun's heat and the way it falls on the Earth's surface – create several zones. The Equator has always been the hottest region, where the combination of heat, evaporation and rain produce the most luxuriant vegetation.

North and south of the Equator, in the so-called tropics, the climate is dominated by the trade winds, with effects that vary with seasons and geography. Seas allow the creation of rains, landmasses blot it up. Central America and northern Australia are well watered; East Africa, sheltered by the bulge of Africa, is often drought-stricken.

Just out of reach of this windy region, between 20° and 30° north and south, little rain falls, creating a desert region – the Sahara, the Kalahari, Central Australia. This is a buffer zone between the unruly tropics and the equally unruly temperate latitudes. Here, daytime temperatures routinely reach 49°C (120°F) and annual rainfall seldom exceeds a few inches.

Farther north and south still lies the most attractive climate in human terms, the zones with the so-called Mediterranean climate. In fact, with today's distribution of continents, such zones are mostly limited to the Northern Hemisphere. Greece, the southern tip of Africa and parts of California exemplify this zone: hot summer days, cool nights and – in the words of the schoolboy mnemonic – warm, wet westerlies in winter.

Northwards and southwards again are the temperate zones, which nowadays lie between 40° and 60° north and 35° and 55° south (the difference being dictated by the preponderance of oceans in the south). Here, westerly winds heavy with rain bring an ever-changing succession of high and low-pressure systems to coastal areas, while the continental heartlands experience far greater extremes.

At higher latitudes, beyond the temperate zones, lie the polar regions. In the northern landmasses, forests give way to tundra; in the south, the oceans become ever colder. Here the winters are long and hard, the summers doing little more than melt the surface of the ice and snow.

EXTREMES OF WEATHER

All the time, the system is constantly tending towards equilibrium – and never achieving it. Each weather system, each blizzard and hurricane, can be seen as a reaction, as the Earth's climatic machinery jars itself out of step and is jarred back towards its ideal. Weather is, in effect, the Earth's atmosphere everlastingly fine-tuning itself, in ways that simply cannot be predicted with any hope of accuracy more than about three days ahead.

Take one of the most dramatic manifestations of bad weather: hurricanes. These vast, turbulent storms are governed by certain rules. They build only over water, and only at certain times of year. A particular combination of heat is required to evaporate enough moisture from the tropical oceans to create a damp column of rising air. As it rises, the air column creates a whirlpool of clouds, wind and rain, fed continually by cooler air sweeping in from below. The whole system, meanwhile, moves westwards and away from the Equator at about 15 mph (24 km/h), driven by the combination of trade winds and the Earth's spin. Worldwide, there will be two or three-dozen hurricanes every year. So far, so good. But there is no way to tell exactly when a hurricane is going to form, how strong it will be, or where exactly it will go before

BLINDING SAND *In desert regions, such as Tunisia, sandstorms are whipped up as cool air rushes in to replace rising hot air.*

BALMY CLIMATE *Patmos, Greece, in high summer typifies the Mediterranean climate that has nurtured European civilisation for over 5000 years.*

it strikes land. There, deprived of its fuel – water – it finally blows itself out.

Today, huge efforts go into tracking hurricanes because the damage they do is so staggering, and prediction saves lives and money. But even now there would be little enough warning if ever there were to be a repeat of the greatest recorded hurricane, the one that hit the Caribbean in 1780. It hit Barbados in October, wheeled north to Bermuda and died in mid-Atlantic. Along the way, it killed some 20 000 people. Barbados's governor recorded the effects: 'Nothing can compare with the terrible devastation that presented itself on all sides; not a building standing; the trees, if not uprooted, deprived of their leaves and branches; the most luxuriant spring changed in this one night to the dreariest winter.' A cannon was blown 140 yd (130 m) along the shattered battlements.

In more vulnerable areas, hurricanes can be far more destructive to human life. The hurricane that hit the lowland plains of Bangladesh in 1970 threw up a tidal wave that flooded 10 000 sq miles (26 000 km²) and drowned unknown numbers of people, perhaps as many as 500 000, with another 2.5 million people succumbing to disease afterwards.

Northern Europe is beyond the reach of hurricanes, except – the weather machine throws up many exceptions – for the one that struck Britain in 1703. The event, which has gone down in history as the Great Storm, was a true hurricane. As recorded by the novelist and journalist Daniel Defoe, it struck on November 26, a Friday. Winds from the south-west built to gale force, peaking in the West Country around midnight. The great Eddystone lighthouse in the English Channel, off Plymouth, was swept away, along with its designer, who happened to be staying there at the time. Wind-driven waters surged up the Bristol Channel,

BARE LANDS *Summer in the Alaskan tundra, too far north for trees to grow, gives the ground a pastel wash of mosses and scanty grass.*

flooding the town and surrounding country, toppling steeples and rolling up church-roof lead 'like cloth'. The eye of the storm passed Liverpool, bringing the winds to London at about 3 am, sending chimneys and roof tiles crashing into the streets and blowing 700 ships together on the Thames in a vast shambles. In all, Defoe estimated that 14 000 houses were destroyed. At sea, 1500 sailors died, and the total loss of life from the storm was 8000.

The storm was also notable for another meteorological phenomenon, recorded by a parson near Besselsleigh, Oxfordshire. He had never seen anything like it before, and had no idea what it was. 'On Friday, the 26th November, in the afternoon, about four of the clock, a country fellow came running to me in great fright, and very earnestly entreated me to go and see a pillar, as he called it, in the air in a field hard by. I went with the fellow: and when I came, found it to be a spout marching directly with the wind. And I can think of nothing I can compare it to better than the trunk of an elephant, which it resembled, only much bigger . . . it crossed a field, and what was very strange . . . meeting with an oak that stood towards the middle of the field, snapped the body of it asunder.'

It was, of course, a tornado, perhaps the most terrifying of all natural phenomena. Though the details are obscure, tornadoes are born when cold, dry air and warm, moist air collide, producing a strong updraft that creates a vicious circle, quite literally: air rising fast and drawing in more air,

which fuels a faster twist and a faster rise, until it reaches the very limits of what air in motion can do. Typically, tornadoes are about 650 ft (200 m) across, travel along at 30 mph (48 km/h) and seldom live for more than a few minutes.

In that time they wreak terrible damage, with winds of up to 300 mph (500 km/h) shrieking around a low-pressure area just a few yards across. If a house withstands the initial impact, the sudden drop in pressure in the tornado's eye can cause the house to explode. Cars blow along like tumbleweed and people are tossed into the air. Tornadoes are quite common: in the USA there are up to 1000 a year, most occurring in 'Tornado Valley' from northern Texas through Oklahoma to Kansas. In the 'Super Outbreak' of 1974, 148 tornadoes killed 315 people across a dozen states. In 1925, tornadoes killed 689 across Missouri, Illinois and Indiana. When a tornado struck Goshen United Methodist Church in Alabama on Palm Sunday, 1994, it lifted the roof off and flattened the walls, killing 20 people and injuring 90 within a few seconds.

BOLT FROM THE DARK
Lightning strikes a sycamore tree, rebalancing the difference in electrical potential between storm cloud and Earth.

One other consequence of thunderstorms is lightning, which could have provided the impetus for the emergence of the first forms of life. Meteorologists estimate that the Earth as a whole produces some 6000 lightning strokes a minute. In human terms, this is enough to cause considerable damage. In

INSIDE A TORNADO

Until recently, tornadoes were not only the most terrifying of natural phenomena, they were also among the most mysterious, pouncing with only a few minutes' warning, tossing cars and people in the air and shattering buildings. All that anyone could say about them was that they were spawned by thunderstorms, and were fuelled by a constant infusion of warm, moist air. But how they worked, no one knew. Nor was there any way to predict why any one storm would turn into a twister.

Then, in 1972, a US research programme, the Tornado Intercept Project, began to gather new data using radar, which allowed forecasters to give better warnings – up to 35 minutes. In the mid-1990s an intensified research project known as VORTEX (Verification of the Origins of Rotation in Tornadoes Experiment) provided some new insights.

Twisters are formed in 'super-cell' storms, cloud structures towering up to 65 500 ft (20 000 m), formed by warm, moist air flowing north from the Gulf of Mexico. Warm air rises, cools and forms water droplets, which sink through warmer air that is still rising. This collision generates the first element in the equation, a thunderstorm. But tornadoes require something more: wind-shear. This happens when ground-hugging winds blow more gently than higher ones. The higher wind rolls over the lower one, generating an invisible roller of air. When this rolling column hits a super-cell updraft, the updraft hauls the rolling column upright, creating a column of air spinning vertically – a 'mesocyclone' several miles wide.

Now the crucial final phase occurs, one that still holds its secrets. A smaller column of air reaches down from the storm cloud and somehow feeds into the spinning column, which contracts and speeds up. Exactly how the downdraft is created and what its role is must await yet more data.

FUNNEL OF DEATH *A tornado, its tight column shrieking with 300 mph (500 km/h) winds, sweeps across the plains of the American Midwest.*

the USA, some 600 people are killed and 1500 injured by lightning every year.

Such unpredictable extremes of the weather occasionally produce truly bizarre events. The oddest are those rains that bring showers of small animals, the product presumably of distant whirlwinds. In the early summer of 1939, a Mr E. Ettles, superintendent of the municipal swimming pool in Trowbridge, Wiltshire, wrote to *The Times* to describe how he had run for shelter from a shower when he heard behind him what sounded like lumps of mud falling. 'I turned, and was amazed to see hundreds of tiny frogs falling onto the concrete path surrounding the bath. It was all over in a few seconds, but there must have been thousands of these tiny frogs, each about the size of the top of one's finger. I swept them up and shovelled them into a bucket.'

Several other frog-showers have been recorded, as well as other oddities. In 1881, several tons of periwinkles and small crabs fell on Worcester after a violent thunderstorm. Some people collected them by the sackful and sold them. Snails rained on

Drought Victim *Lake Bildon in Western Australia is sucked nearly dry as a result of the sporadic rainfall and high temperatures of the outback.*

Redruth in 1886, worms on Clifton, Indiana, in 1892, fish on Bareilly, India, in 1893. There have been reports of hailstones weighing 2 lb (1 kg), and hailstorms that slew birds by the ten thousand.

The Cycle of Water

One vital feature of the Earth's restlessness is the way in which water is recycled. It is the essence of life, and the major force that remoulds the Earth's surface. Without the constant turnover of water through the process of evaporation from the oceans and its deposition as rain and snow on land, no higher form of life could ever have developed, no mountain would be eroded, no river carry nutrients from land to sea.

Most of the water vapour in the atmosphere – about 84 per cent – comes from the oceans, particularly from equatorial waters. Once in the air, the vapour is carried about by winds, condensing back into ice and water droplets and falling as snow, rain and hail. Most of it falls back into the seas.

Only a little falls on land. The total volume of water on Earth is roughly 330 million cubic miles (1320 million km³), of which 97 per cent is in the oceans. Only 3 per cent is on the land, and 75 per cent of that is locked up in ice in the polar regions and in glaciers. Almost all of the remaining fresh water lies below ground. All the lakes and rivers of the world, all of the wetlands and damp soils, account for less than 1 per cent of the total, and a mere 0.035 per cent is in the air at any one time. Yet this is enough to give the land an average of 35-40 in (90-100 cm) a year in rainfall, if it were evenly distributed.

Currents and Tides

A final major component in the Earth's machinery of climate and weather is the circulation of water by oceanic currents. The oceans have always been stirred by currents, powered by a combination of temperature differences and the Earth's spin, though their patterns must have changed many times as the continents shunted about the surface of the Earth.

Cooler waters near the poles are denser, and therefore heavier. The top layers, where ice forms, is cooler still and sinks, flowing underneath warmer seas to the north and south. The Earth's spin focuses these flows into slow-moving, deep currents, while prevailing polar winds create a powerful surface current. Since most of the Southern Hemisphere is open ocean, this effect is most obvious in the Antarctic, where the westerlies create a South Polar Drift that drives currents farther north.

Trade winds also create and amplify currents, building steady drifts from east to west across both the Atlantic and the Pacific. When these currents meet continents, they are turned slowly around, creating other currents, such as the Gulf Stream, which is responsible for moderating the otherwise severe climates of northern Europe.

None of these processes operate independently. As an example, a Pacific current that flows down the coast of South America has been identified as the cause of dramatic changes in weather systems, in a way that links the mechanisms of the Earth's spin, currents and airflows. The current has been named El Niño, 'the Child', by Peruvian fishermen because it appears off the Peruvian coast just after Christmas. Most years, it flows for a few weeks, overriding the deeper, colder, regular currents that bring nutrients north from the Antarctic. Fish disappear and prevailing winds change.

DEATH BLAST *Near Mount St Helens a shattered tree stump emerges from volcanic ash, a victim of the explosion that destroyed the mountain in 1990.*

Every few years – anything from three to seven – El Niño is warmer and stronger than usual, bringing additional masses of warm water, and can last for as long as a year or two. Fish die by the million. This is not a merely local event, but is caused by changes thousands of miles away, in the western Pacific. There, normally, the trade winds pile up the warmer waters on the Pacific's surface in the far west – towards South-east Asia – where tropical storms are brewed. When the trade winds weaken, they allow the warm waters to spread back eastwards towards South America, bringing the storms with them. The changing weather patterns, known as the Southern Oscillation, that create an abnormal El Niño event affect weather over a vast area. In 1982-3, it caused drought in South-east Asia; forest fires in Indonesia; withered crops from the Philippines to Botswana; hurricanes in Polynesia; and floods in Peru and Ecuador. In 1994, similar changes kick-started drought in Australia, bringing ruin to farmers and destruction by fire for millions of acres of forest.

Another variable in the evolving Earth is the twice-daily tidal flow induced by the Moon and the Sun. What role the tides played in the evolution of life is unknown, but certainly they became vital when early life forms began to colonise the land some 500 million years ago. Back then, when the Moon was closer to the Earth and days were half the length of today's, the tides were twice the size, and rose and fell at twice the speed. If today's tides blur the edges of the land, rising and falling anything from 15 to 50 ft (4.5-15 m) in about 12 hours, imagine tides that roar in and out every three hours, averaging 50 ft (15 m) and sometimes exceeding 100 ft (30 m). In low-lying areas they would have exposed many square miles of rock, sand and mud twice in the much shorter days, opening huge areas for newly evolving life forms to exploit along the frontier between sea and land.

VOLCANOES

Forced up by the Earth's internal fires, molten rock – magma – continually probes for outlets, finding them through weak points and vents, mostly along the edge of the continental plates. The molten rock, ash and gas emerge as volcanic eruptions, the quick-fire forges of the changing world. They build mountains, create new rocks, spread ash over huge distances, re-form the atmosphere, change weather patterns and create new islands. When the oceans formed, there was much more volcanic activity, but today's volcanoes give an indication of the effects. The Earth has some 500 active volcanoes, with 50 eruptions annually, about 70 per cent of them on the 30 000 mile (48 000 km) rim of the Pacific plate, the so-called Ring of Fire.

By any standard – local or global, over time scales of seconds or millennia – the outbursts are often catastrophic. In 1902, the pretty town of St Pierre on Martinique – the 'Paris of the West Indies' – was struck by an explosion from nearby Mont Pelee that killed 30 000 people in four minutes and blew the whole place to pieces with the force of a nuclear blast. One of the greatest eruptions of all time was the explosion of the Mediterranean island of Thera, now known as Santorini, in 1450 BC. Some four to five times the power of Krakatau, the explosion blew apart an island that was 10 miles (16 km) across, leaving a mere crescent of rock. The resultant tidal wave and cloud of ash shattered the Minoan civilisation on Crete, 80 miles (130 km) away.

Eruptions like this re-create landscapes and alter climates. And recorded eruptions are nothing compared to the infrequent ones that have punctuated Earth's history. Some 600 000 years ago, an eruption in what is now California's Yellowstone National Park threw out 240 cu miles (960 km³) of ash, enough to bury London a mile (1.6 km) deep, remodel the land for hundreds of square miles, darken skies around the world, and turn summer to winter.

THE RESTLESS EARTH

Within the constraints imposed by physical and chemical laws, processes evolved that would mould the Earth, providing it over billions of years with a huge range of slowly changing features, systems and character traits.

Winds, weather, ocean currents and tides, in conjunction with the slow shifting of the continental plates and slight irregularities in the Earth's orbit, were the systems that came to dominate the Earth's surface some 3.8 billion years ago. With these mechanisms in mind, consider the Earth again around the time that oxygen re-formed its atmosphere about 2.5 billion years ago.

The following account focuses on the major changes, and compresses vast vistas of time. In human terms, the units are so huge as to be almost incomprehensible. Human-like creatures have been on Earth for only 0.0006 per cent of its history: in the famous image that divides Earth's history into a day, all of human history has lasted a mere eyeblink. Even if the Earth's history is compressed into a year, human history would still last only 30 seconds. But to understand the context for those 30 seconds demands an understanding of the whole year. In Earth's life story, 10 million years is a good base unit, long enough for a mountain chain to rise or fall, long enough for a continental plate, moving half the length of your little finger every year, to migrate several hundred miles.

What will emerge from this dramatic compression of time is the Earth's dynamism. Rather as a flower's unfolding is intensified by a speeded-up film, a fast-forward view of Earth's history shows just how active it is. It was Professor James Lovelock who coined the term Gaia to describe the Earth, using the name of the classical Earth goddess to convey his sense that the whole planet was in some way alive.

ARCHEAN TIMES

The 1.5 billion years that preceded the emergence of an oxygen-rich atmosphere 2.5 billion years ago forms a period now known as the Archean. Until the 1950s, this distant time was simply seen as part of the Precambrian, a hidden precursor to the time when life was *thought* to have begun some 550 million years ago, according to research mainly done on rocks in Wales (Cymru in Celtic, or Cambria in its Latinised form). Then minute fossils were found in Precambrian rocks, and dating techniques improved. Since the 1950s, scientists have added two chapters of Precambrian history, the Archean followed by the Proterozoic ('former life'), the boundary between them being drawn approximately 2.5 billion years ago. These eras provide vital foundations for the emergence of life.

Archean rocks can be dated only by measuring the decay of certain radioactive elements. Since most of the rock has been eroded away and much of what is left is hidden by later formations and earth movements, the Archean is largely a dark age. One third of Earth's history is still hidden from us, but at least a general picture has emerged. For this immense stretch of time, the surface was torn and twisted by subterranean forces. From volcanic mountains and rifts, molten rock poured out to create barren tablelands. These so-called 'cratons' would form the hard cores of the future continents – enduring, rocky rafts borne about the world's surface millimetre by millimetre by the currents in the underlying mantle. Then as now, these currents hit the surface in mid-ocean cracks, from which emerged new magma, the molten matter formed in the Earth's upper mantle, rejuvenating the ocean-bed rocks, forcing older rock aside, and creating the rocky travelators that bore the continental cores along.

The oceans were chemical soups, rich in detritus washed from the rocky shores and rich also in microscopic forms of life, single-celled organisms and the blue-green

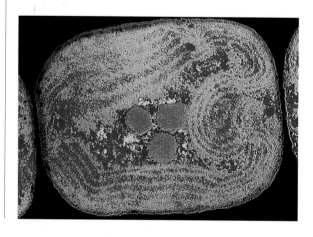

REVOLUTIONARY CELL
Single-celled blue-green algae injected oxygen into the Earth's early atmosphere until it became oxygen-rich. Mauna Loa, Hawaii (opposite), is one of hundreds of volcanoes that contribute vast quantities of carbon dioxide to the atmosphere today.

algae that were transferring oxygen from the rocks and water into the atmosphere. Already the processes of erosion and deposition had laid down sedimentary rocks – sandstones and shales, limestones and chalks – while on land continental drift had forced up mountain ranges, from which rivers carried sediments to the seas.

What the geography of the Earth's surface looked like at this time is anyone's guess, because so little of the ancient rock has been identified. All anyone can deduce from the little patchworks that emerge on the surface is that major landmasses were in existence, but their shapes and positions would have borne no resemblance to today's world. There were probably about two dozen cratons, which fused into supercontinents and separated into islands many times. Two parts of the Canadian shield were apparently welded together permanently around 2 billion years ago, creating a range of mountains between the Great Slave Lake and Lake Winnipeg. The range vanished long ago, scrubbed flat by 1 billion years of rain and ice. Today the area is all scanty fir forest and small lakes.

WORLDS OF ICE

Around this time, another of Earth's fundamental traits emerged: ice, in overwhelming amounts. However, scientists had first to see the significance of ice as a tool in the formation of the modern world before they

could understand its significance in the remote past. That quest raised the problem of why ice comes and goes, and the answer links two other themes – the 19th-century controversy over the age of the Earth and the peculiar irregularities in the Earth's orbit and spin analysed by the Serbian, Milutin Milankovitch.

The idea that much of the world was once covered with ice is now commonplace. When it was first voiced, however, it was as controversial as the theory of continental drift; and once accepted, just as revolutionary. In hindsight, the idea seems self-evidently true, but not in 1837. At that time, the world of learning in the West was gripped by arguments between the catastrophists – those who explained geological

FOREIGN ROCK *Glacial erratics are found in County Clare, Ireland. Brought by advancing glaciers, the boulders were dumped when the ice melted.*

forces in terms of cataclysms, in particular the Biblical Flood – and uniformitarians, who saw all geological forms as the product of infinitely slow processes extending over millions of years.

One particular problem that gripped geologists was that of the so-called 'erratics', rocks that had for some reason been transported from their points of origin, which might be a few miles away, or in some cases hundreds of miles away. Catastrophists had a ready answer – the rocks had either been carried there by the fierce onslaught of the Flood, or they had been captured in icebergs and released in their final resting places when the icebergs melted. Some of the arguments supporting these ideas now seem naive. Scientists in Cambridge spent a good deal of time and effort analysing the supposed properties of 'waves of translation', as they were known, waves of such magnitude that they could not possibly exist in the present-day world. Other arguments were more convincing – on the voyage of HMS

THE LITTLE ICE AGE

Within the long-term trends that govern major ice ages are minor fluctuations, probably caused by changes in the upper-atmosphere jetstream. One such fluctuation was the so-called Little Ice Age, which started in about 1300, peaked around 1450 and ended in the mid-19th century. The effects entered folklore – establishing the ideal Christmas as white – and influenced history.

Before the onset of the Little Ice Age, the Vikings were free to explore

northern sea routes to America and grapes flourished in southern Britain. As the cold spread, however, Viking settlements in Greenland failed and Britain lost its wines until the 20th century. Glaciers inched out of Alpine valleys. Colonists in America endured winters of legendary harshness, the American Midwest was hit by drought, and Dutch painters recorded skaters on the canals.

Before about 1400, the River Thames had frozen over only about

once every 100 winters. Between 1407 and 1814, it froze about every ten years, inspiring 'Frost Fairs' in the 17th century, with tents erected on the ice and horses and carts driven across it. On February 2, 1814, a circus elephant was walked across the river at Blackfriars, as reported in *The Times*: 'The singularity of such an animal on the ice attracted a great concourse.' Four days later the ice broke, and the Thames has not frozen since. The Little Ice Age was over.

Beagle, Charles Darwin himself saw icebergs that contained boulders. Yet the erratics remained mysterious. How on earth could waves or icebergs account for erratics found 5000 ft (1500 m) up mountains hundreds of miles from the sea? Where could all that water have come from? And where could it have gone?

There were already those who suspected the truth. They were, naturally enough, those scientists who knew most about the behaviour of ice – the Swiss. In 1815, a chamois-hunter named Jean-

THE POWER OF ICE *Glaciers like the Brady in south-east Alaska are reminders of the ice sheets that once spread across northern America and Eurasia.*

Pierre Perraudin concluded that a glacier near his home in the southern Swiss Alps had once filled the whole 24 mile (38 km) long valley. The idea was picked up and extended by Jean de Charpentier, director of salt mines in Bex, and extended again by an

ICY FRONTIER *Ice breaks up in the ocean around Antarctica, creating cold currents that affect sea and land environments far to the north.*

eager young geologist named Louis Agassiz.

Agassiz expounded his theory in a lecture to the Swiss Society of Natural Sciences in Neuchatel on July 24, 1837. What he said left the society's members astonished and outraged. From the existence of erratics and scratch marks on rocks, he argued that the mountains roundabout – the Jura – had once been covered by ice. Moreover, this sheet was part of an immense polar ice sheet that

ICE-AGE REMNANT *These ridges,
or eskers, in North Dakota are
the eroded sediments deposited
by rivers that cut through or
under long-vanished ice sheets.*

all organic life at the Earth's surface . . .
The silence of death followed . . . springs
dried up, streams ceased to flow, and sun-
rays rising over that frozen shore . . . were
met only by the whistling of northern winds
and the rumbling of the crevasses as they
opened across the surface of that huge
ocean of ice.'

FROM THEORY TO FACT

It took time for the theory to win accep-
tance, but established geologists in Europe
slowly came to agree with him. Americans
proved more open to the new ideas, on the
other hand, and Agassiz moved to the USA,
where he became a professor at Harvard.
American geologists identified exactly
where the Ice-Age glaciers had ground to a
halt: all of Long Island, it appeared, was a
'terminal moraine' piled up by the leading
edge of an ice-sheet that had been some
5000 ft (1500 m) thick. After some 30 years,
the existence of an ice age had become the
new orthodoxy.

This begged many questions: what was
the effect of such a change? And if one ice
age, why not several? And anyway: *why?*

One effect of the Ice Age – besides the
impact on plants and animals – was almost
immediately apparent. To make an ice
sheet 1 mile (1.6 km) thick takes a great
deal of water, thousands of cubic miles of it.
The only source of such immense amounts
of water was the sea. A simple sum produced a dramatic conclusion: the coming
of the Ice Age lowered the sea level by some
350 ft (100 m), changing the geography of
the world. During the Ice Age, the West Indies would have been united, the English
Channel empty, Alaska would have been
linked to Siberia by dry land, and every
landmass would have increased in size.

The great weight of ice explained other
mysteries. One mystery was that posed by
the existence of seaside features showing
that the sea level had been much *higher* immediately after the Ice Age. In Scandinavia,
for instance, recent marine deposits are
found at over 1000 ft (300 m). If the ice had
locked up so much water, and had not yet
fully retreated, how could there have been
such a flood?

covered all Europe down to the Mediterranean. Europe had, in his words, been in
the grip of an 'Ice Age':

'The development of these huge ice
sheets must have led to the destruction of

In 1865 a Scottish geologist, Thomas Jamieson, came up with the answer. It demands a comparison of ice with rock and rock with ice. In terms of weight, an ice sheet acts as an extra mountain range. And mountains themselves are weighty things, while the mantle beneath them is relatively soft. The result is that mountains, like icebergs, sink until they are in balance with their surrounding rocks and the underlying mantle. Any ice-covered continent – in this case all of Europe, Asia and North America – would have been depressed until the ice melted. At this point, the landmasses lay anything up to 1000 ft (300 m) below their ice-free level. Released, they slowly returned to their 'normal' position, carrying their shorelines and marine deposits up with them.

It makes for a complicated picture: sea levels falling and rising as ice forms and melts, land levels altering too. But the two processes do not work simultaneously. Rock is a slow-moving medium and takes millions of years to react to the relatively sudden imposition and removal of ice.

The presence of ice sheets explained other features of the world. For instance, a million square miles of Europe, Asia and North America are blanketed by sedimentary rocks known by the German term, *loess*. These deposits can be of immense thickness, as anyone who has seen the bluffs in the Mississippi valley knows. Some loess cliffs in China are 500 ft (150 m) high. But they are puzzling because, although obviously made under water, they contain no marine fossils and no sign of being formed under water, since they have no stratification.

Again, ice provided the answer. As explained by a German geologist, Ferdinand von Richthofen, in 1870, the melting ice-sheets left great banks of silt. The silt was not frozen in place or anchored by plants. The result? It blew away. South of the retreating ice, great dust storms scattered the silt until it lay dozens or hundreds of feet thick, creating not only loess rock but also the rich soils of America's farming belt.

Once this idea had taken root, the explanatory power of the theory became clear. Glaciers could act as giant files, carrying rocks in their bases and grinding away at the land beneath. The scratch marks survive in countless rocky outcrops. Valleys and mountains were formed and re-formed by ice and meltwaters, rocks transported hundreds of miles, soils mixed, rivers made, landscapes stripped – Canada's rich farmlands are fringed to the north by landscapes scrubbed, shaved, filed and polished by ice, leaving shallow soils and small lakes.

Almost as soon as geologists began to explore the once-glaciated lands, they found several different sequences of glacial deposits and meltwater layers. They realised that the Ice Age was not a single age at all. There had been a succession of glaciations – four, in fact. Since the ice sheets, the glacial debris, the meltwaters and the scattering of the loess soils all overlaid each other, the dating of these four episodes proved extremely difficult. It took the best part of a century.

Only in the 1960s, after teasing conflicting results from a variety of sources – including layers of rock, fossil remains, cores of sediments taken from the ocean floor, radiocarbon dating – did researchers stumble on the pieces of evidence,

GLACIAL POLISH *These basalt rocks in the Devil's Postpile National Monument, California, were polished smooth by a glacier.*

in two sites that could hardly have been more different. The first was a brick quarry near Brno in modern Slovakia, where several different layers of loess soils preserved a good record of the magnetic reversals. The other site was at the bottom of the sea. Cores drilled from the Caribbean provided a record of past temperatures, revealed by a somewhat arcane technique: measuring the amounts of an isotope of oxygen that was removed by ice and replaced by meltwater.

The results matched, and in the early 1970s for the first time scientists could agree on a chronology for the last four ice ages. Over the past 500 000 years, ice ages were spaced about 100 000 years apart, developed slowly over some 25 000 years, and were terminated over about 10 000 years. But it took a long time for the ice sheets to retreat completely, and for sea levels to return to normal.

MILANKOVITCH'S CYCLES

Meanwhile, of course, the search was on for the cause. Why did the ice come when it did? Why did it retreat again? The answer takes us back to the details of the Earth's orbit analysed by Milankovitch. His life's

work was to measure the inflow of heat from the Sun as modified by the tripartite variations in the Earth's motion – the 100 000 year cycle as the orbit became more elliptical and more circular; the 41 000 year cycle as the tilt varied; and the 22 000 year cycle dictated by precession. He was convinced that these variations were enough to alter the Earth's heat budget to the extent that it would kick-start ice ages, which then became self-perpetuating as increasing areas of ice and snow reflected ever greater amounts of heat back into space. Only when the cycles combined to pour additional heat onto the Earth's surface did the mechanism flick into reverse.

Throughout his life, and long after his death, Milankovitch's theories remained controversial, because the effect of the cycles was so hard to separate from the 'noise' of other influences, like changes in ocean currents and the time it took for the sheets to form and melt. Only in 1976 did scientists become convinced that the varying motions of the Earth round the Sun triggered ice ages. But if the history of the last 500 000 years was defined by a succession of ice

ages, and if those ages were brought about by the steady beats of the Earth's spin and orbit, surely that should mean that the Earth had always been subject to ice ages?

This, however, is not so. Ice builds on land more than on water. The Antarctic is an ice cap, but the Arctic is not. So there is another crucial element to be taken into account: the position of the landmasses. Moreover, shifts in the position of landmasses meant shifts in ocean currents, and thus shifts in the distribution of heat. Some primeval Gulf Stream Drift heading due north or south would be enough to keep otherwise frozen lands ice free.

And now it is possible to return to the Earth's history some 2.5 billion years ago – with one proviso. Occasionally, and randomly, the Earth was struck by asteroids and meteors, a continuation of the bombardment that had been such a feature of its earlier life. These large-scale impacts would have had immense effects: tidal waves hundreds of feet high, holes punched right through the Earth's crust, dust clouds far more pervasive than any caused by a volcano.

When they struck, where they struck and what damage they did are still highly controversial issues, but scientists now agree on at least one massive strike – the one that contributed to the extinction of the dinosaurs 65 million years ago. Where there has been one, there must have been others. Though geological change will have eradicated almost all easily accessible evidence, scientists hunting ancient craters cast suspicious eyes on great 'unnatural' geographical curves, such as the bulge of China and the Nastapoka 'curve' in Hudson Bay. The effects of such massive impacts would have included the extinction of numerous species; and also, perhaps, the initiation of an ice age, by blanketing the Earth in a cloud of ash. Once kick-started, increasing cold might well overwhelm Milankovitch's cycles.

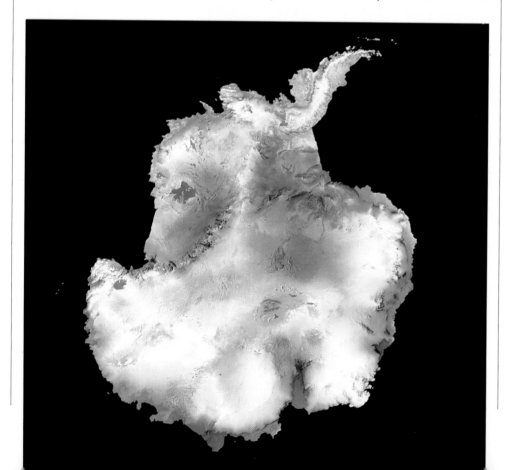

ICY BLANKET *A false-colour view of the Antarctic reveals the mountains (red areas) that are mostly hidden beneath the 2 mile (3 km) thick ice sheet.*

THE HAWAIIAN CHAIN: A VOLCANIC CONVEYOR BELT

The half-dozen main islands that make up Hawaii today – in particular the low-lying volcano of Kilauea on Hawaii island itself – provide evidence of a deep-seated geological trait. They are the products of a 'hot-spot', a plume of molten rock surging up from the depths of the Earth. This hot-spot has been a semipermanent feature of the Earth's core and mantle for hundreds of millions of years. But the Pacific plate that caps it is no more static than any other plate. It overrides the hot spot, which steadily cuts through it like an acetylene torch, only to be blocked off again as new rock slides into place. The result is a chain of volcanic islands – not only those that form Hawaii today, but a whole assembly line of 17 islands and island groups stretching 1500 miles (2400 km) north-west to Midway Island.

FIERY SIGNPOST
The Hawaiian volcano
Kilauea is one of a line that
marks a weak, island-building
hot spot in the Earth's crust.

With this random and unquantifiable element in mind, let us see how the Earth's changing geography defined the world we see today.

THE CHANGING FACE OF THE EARTH

Ice came early to the cooling Earth. As the continents have been shunted all over the Earth's surface since the first appearance of ice, and as most of their rocks have been recycled, there are precious few clues as to the geography of the planet then. But something has emerged from those remote times. In Ontario in southern Canada, and several sites farther north, signs of heavy glaciation – a muddle of boulders, pebbles, clay and sand mixed up with deeply scratched rocks – date from about 2.2 billion years ago. There are also the remains of glacial lakes, which deposited thin layers of mud, now finely laminated mudstones. Other signs of glaciers of the same age have emerged in Australia and southern Africa. So southern Canada was ruled by Arctic conditions then, as now, and must have been near a pole – quite possibly the South Pole, with Africa as a near neighbour.

By about 1 billion years ago, all the major landmasses seem to have been swept together, forming a supercontinent lodged firmly over the South Pole with one huge peninsula stretching beyond the Equator. This arrangement seems to have triggered the longest cold period in the history of the Earth, an ice age lasting 100 million years that spread its ice tentacles to within a few degrees of the Equator. Today, glacial deposits of this age are dotted across all the Earth's landmasses except India and Southeast Asia. Perhaps back then this region, though barren of life except along its fringes, was a tropical paradise compared to the frozen wastes farther south. Or perhaps the glacial evidence, if it exists there, has still to be found.

Take another immense step forward in time – 500 million years, to the time geologists call the Cambrian. On the 24 hour clock of the Earth's history, 21 hours have already passed, about which we have little evidence. Now, in the Cambrian period,

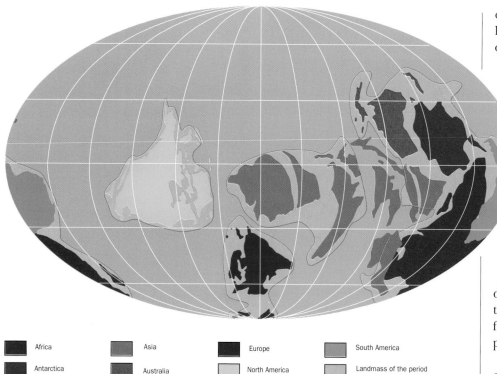

Africa	Asia	Europe	South America
Antarctica	Australia	North America	Landmass of the period

SHIFTING CONTINENTS *Around 500 million years ago the main landmasses were Laurentia (North America), Baltica (Europe), Angaraland (Siberia) and Gondwanaland (Australia, Africa, Antarctica and South America).*

evidence comes crowding in, mainly in the form of fossils of shell-covered marine animals. The continents had broken apart and wandered into the tropics. The glaciers had melted and the sea risen, reducing much of the land to huge islands between which the ocean currents were free to circulate.

Similarities in surviving rock formations allow geologists to glimpse the identity of present-day landmasses in these islands, and give them names. North America and Greenland formed Laurentia, which was separated by the Iapetus Ocean, a precursor of the Atlantic, from Baltica, which grouped together bits of present-day northern Europe. Siberia was off on its own, part of a free-ranging island known to geologists as Angaraland. Straddling the Equator and reaching down towards the South Pole was a supercontinent, Gondwanaland, linking

all today's other major landmasses – Africa, Arabia, India, Australia, Antarctica. At that time, the coldest part was the future West Africa, while the bit that eventually became Antarctica was parked on the Equator.

Fifty million years later, the Iapetus Ocean was smaller. Laurentia and Baltica – North America and northern Europe – were being driven together, with consequences that help explain why the north-eastern United States looks the way it does.

First, for many millions of years before any life evolved, the great island-continent

of Laurentia drained eastwards into a shallow sea that ran from Labrador to the Gulf of Mexico (not that the coastline was anything like today's). As Baltica was forced into and over Laurentia's low-lying coast, the sediments buckled. Great mountain chains rose, as jagged as the Alps and Himalayas, and probably as high, forming peaks of 20 000 ft (6000 m) and more. These heights are guesswork, for even as the mountains formed, water, frost and gravity assaulted them. Rock was ground to gravel, sand and clay. By some 500 million years ago, the mountains were mere hills. Now, they are hardly there at all. If you drive through lower New York State along the Taconic Parkway, you can see the stubby, forested remnants of the once-towering peaks, the Taconics.

Other remnants of these slow-motion continental collisions are the mountains of northern Britain, eastern Greenland and western Scandinavia. These ranges were created during what is known as the Caledonian Orogeny (mountain formation), a massive and complex 140-million-year process named after the Roman term for Scotland. Before that time, Scotland was part of North America. When it ground into the rest of Britain, the crash threw up the Grampians.

Farther south, Gondwanaland was swinging sideways, until some 400 million years ago it collided with North America, extending the eastern mountain chains into a continuous belt from Newfoundland to Tennessee, and stitching Florida into its present position. As it swung, the southern supercontinent opened a new sea, the Tethys.

By now, plants had begun to colonise the land, and rapidly adapted to exploit this vast

SEA COLONISERS *Clam-like brachiopods dominated marine life when the first molluscs evolved during the Cambrian, 550-500 million years ago.*

new resource. For over 100 million years, trees and ferns coated the land, laying down the thick deposits of plant residue that one of the Earth's future life forms would discover to be an excellent fuel. Hence the name given to this period, the Carboniferous. For much of this time the Earth was warm, and shallow tropical seas washed over Laurentia and Baltica (North America and northern Europe), leaving huge sheets of limestone. In the tropical north, vast swamps formed until, once again, changes in orbit, spin and ocean currents tipped the Earth into another ice age, sucking the seas from the land.

Meanwhile, the drifting continents had again reformed the face of the Earth. Gondwanaland and Laurentia touched and fused. Where they met – the joined mass of ancient Africa and South America running headlong into the eastern edge of North America – they pushed up the Appalachian system, mountains that pick up from the more ancient Caledonian deformations to the north.

As in any collision, the impact did more than crumple a wing and a bonnet. It ripped off a piece of the other vehicle, in this case Africa. From South Carolina to New Jersey, the underlying rocks match those that fringe the Sahara. These massive changes, stretching over millions of years, had other effects as well. The wounded crust poured out volcanic magma and hot water rich in minerals such as copper, zinc and lead. These reacted with the organic matter in the more ancient limestone and produced a range of metal ores that today form the basis of the Mississippi valley's rich mining industry.

There was no Atlantic, and a few scattered islands that would later form central Europe, Bohemia and Siberia were drifting off by themselves. Nonetheless, for the first time the terrestrial world began to acquire a semblance of its modern self: the northern and southern landmasses were roughly aligned, North America and Europe in the north, South America, Africa and Arabia, Antarctica and Australia in the south.

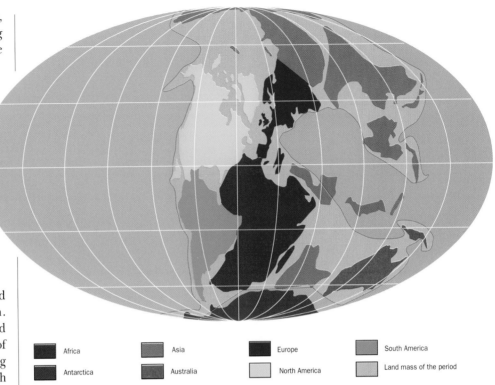

Africa	Asia	Europe	South America	
Antarctica	Australia	North America	Land mass of the period	

The period during which this occurred, the Permian, is named after a series of sandstone rocks first identified in 1841 near the town of Perm, on the European side of the Urals. Up until then, the combination of trade winds and the Earth's spin would have ensured the steady flow of a tropical current between the two landmasses. Almost in an instant – just a few million years – the collision created a dam. The current would have been deflected north and south, and thousands of miles of coastline vanished. Since the south of the new supercontinent spanned the Southern Hemisphere, from across the Equator to the South Pole, the new Gondwanaland, with its vast, landlocked heartland far from warming currents, acted as a giant refrigerator, producing effects that geologists have pieced together by examining ancient shorelines and analysing the marks made by ice on rocks.

From about 340 million years ago, the land that later became South America, southern Africa, India and Australia began to ice over. The Sun's heat, mirrored back into space by the ice, was not enough to reverse the trend, and the world plunged

UNITED WORLD *By about 250 million years ago, the major landmasses had swung round into their present hemispheres and fused into one supercontinent.*

steadily into its deepest ice age. Some 65 million years later the south was coated in an ice sheet as thick as Antarctica's and three times the area. The spreading ice sucked up ever more of the sea, until the sea level was some 300 ft (90 m) lower than it is today, and so it remained for some 30 million years.

PANGAEA, AND ITS BREAKUP

About 250 million years ago, there occurred another defining moment in Earth's recent history, when Angaraland, the 2500 mile (4000 km) wide island that included Siberia, butted up against the other united landmasses. The crumple zone where they joined together is still there in the 1250 mile (2000 km) line of eroded stumps of a once-great mountain range, the Urals. Originally, these peaks, now no more than 5000 ft (1500 m) high, would have been jagged young mountains with ranges of active

PRIMORDIAL MONUMENTS *The buttes of Monument Valley, Arizona, are the remnants of sandstone strata laid down in the rainshadow of the young Appalachians.*

volcanoes. Lava piled up and huge rivers contributed to what had become a chemical cauldron, cooking up over 1000 minerals that today are Russia's natural treasure trove. The volcanoes and rivers also formed the wastes of central Siberia, now vast forest and tundra.

Central Asia had joined Europe, and all the world's land had become a single supercontinent, referred to as Pangaea ('land everywhere'). Pangaea consisted of Gondwanaland and Laurasia (Europe, Asia, North America and Greenland). Only a few scattered islands – bits of China and Southeast Asia – remained apart. The world's geography was utterly different from today's world, but it was from this single landmass

that today's emerged. Though the time scales are vastly extended in human terms, in Earth's own terms we are now in modern times. Over 90 per cent of Earth's history has passed, with interesting consequences for geological evidence. The formation of Pangaea is the geological equivalent of the ending of the Dark Ages and the invention of printing. By comparison with the more ancient past, the evidence of Earth's history from this time emerges with startling clarity in the pages of the rocks.

By now, life was well established and had colonised the land, producing the first mammals and a range of reptiles – the dinosaurs and their marine and airborne contemporaries – with a variety that matched that on Earth today. Though apparently locked into one stable landmass, the continents were still under stress, still in motion. Pangaea rotated clockwise, with its leading southerly edges – current South America – grinding into what is today's Pacific, and

throwing up the beginnings of the Andes. There were no Rocky Mountains then, but the Appalachians were still towering peaks, and they caused a rainshadow that created a vast desert over what is now North America. Winds scoured the surface, scattering sand over the interior. In places, those deserts overlap present-day ones, like the Navaho Desert of Utah, formed much later by a reverse process – the creation of the Rockies, with their rainshadow. The awesome stacks and buttes of Arizona are, as it were, the exposed bones of that ancient sandstone, eroded by wind and rain.

Beneath Pangaea, the slow-moving currents of rock, surging up from deep in the Earth, built up pressures that could not, in the end, be resisted. The ancient continental shields remained intact, as always, so the land unzipped along the lines of least resistance, the new young mountainous edges of the continents. About 210 million years ago, Pangaea began to split. Forced apart by molten rock welling up from below and flowing away to either side, Laurasia began to fold away from Gondwanaland, separated by an inlet, which became a gulf, and which would eventually become the Atlantic Ocean.

The face of the Earth split and split again, with effects that can be imagined by looking at the map today. For instance, at one time, perhaps 150 million years ago, a great river flowed south and west from the highlands that joined ancient West Africa to ancient South America. The river flowed west, into what became the Pacific. Two things then happened. West Africa and South America split apart, along a line that would later become the South Atlantic. And South America, drifting westwards, overrode the Pacific plate. Like an ice floe sliding on top of its neighbour, the whole of South America tipped, and a vast mountain range – the Andes – was thrown up along its leading edge. Now there was no outlet for the river, and in any event the lie of the land had changed. The river reversed course. New waterways evolved, running from the Andes eastwards, and the Amazon was born.

In the north, a wing of Laurasia that by now was straddling the North Pole pivoted clear, swinging round across the pole to collide with Siberia, throwing up a line of Arctic mountains and opening the North Pole to the ocean – but also almost damming it in, condemning it to become the frozen ocean it has remained ever since.

To the south, the breakup gathered strength. Between 120 and 65 million years ago, India, Antarctica and Australia were all unzipped along ancient fault lines. Gondwanaland, separated now from the northern

landmasses, was no more. With all its land broken up into a dozen major and several minor segments, and the ocean currents now free to bathe them all, warming continued. The ice caps shrank and the sea rose, drowning low-lying areas, covering huge areas – including all of England – with inland seas. In these clear, tranquil waters, tiny single-celled creatures called coccoliths thrived and died, raining down to form chalk. The results are all around in the chalk downs of southern England, picturesquely exposed by later erosion in the White Cliffs of Dover.

By some 50 million years ago, the world's geography was at last recognisably modern. A gulf still separated Africa and India from Asia and Europe, and a great inland sea, the Tethys, washed over what is now the heart of central Asia. As these plates approached each other the sea floor

THE ORIGIN OF ARABIAN OIL

Of the many slow-motion changes that occurred in Earth's history, one has had particular significance for the modern world – the one that produced the massive Middle East oil deposits.

To make oil takes a peculiar set of conditions. A warm ocean, a rich ecology of minute marine organisms, aeon upon aeon of oceanic life raining down to form organically rich layers of rock, two continental plates meeting each other – but not so violently as to destroy the rocks beneath. When overlain with later

formations, subjected to the right amounts of heat and pressure, then trapped by a bridge of impermeable rocks, the marine sediments turn to petroleum products – gas and oil of many different kinds.

Some 250 million years ago, the northern fringe of Gondwanaland produced just the right conditions, and did so over perhaps 200 million years. About 150 million years ago, the oil was there. Finally, 25-15 million years ago, as Saudi Arabia butted into Iran, the impact threw up the Zagros Mountains. The oil-rich

sediments were folded into a switchback, with the lighter-weight oil and gas trapped in the peaks. In the simplest cases, a pipe inserted into pockets of oil and gas acts like a syringe lancing a boil: the organic matter surges out, as it has done from the oil fields of the Saudi Arabian peninsula for the past 50 years.

FOSSIL FUEL *A flare releases gas pressure in an oil field, part of a process exploiting the 150-million-year-old remains of marine organisms.*

that of Mount St Helen's in 1980, and in the rainshadow that has helped to turn parts of the Midwest into desert.

THE NEW ICE AGE

The Earth was, as always, precariously balanced on the edge of a runaway ice age. What had made the difference – what masked the influence of the cycles identified by Milankovitch – had been the massive, universal landmass of Gondwanaland lodged over the South Pole. Now, although the landmasses were divided and warmed by free-flowing currents, the poles were still not unencumbered. The South Pole was occupied by Antarctica and isolated by an icy circumpolar current, while the North Pole was almost entirely hemmed in by land.

Although small, the changes in heat brought on by Milankovitch's cycles were enough to keep the Earth fluctuating a degree or two either side of a global mean temperature of 25°C (77°F). The cycles interacted, with long-term low temperatures occurring about once every 100 000 years, corresponding to the changes in orbital eccentricity. Other, lesser lows, dictated by precession and tilt, occurred every 43 000 and 24 000 years, with a final minor cycle of 19 000 years. The interaction between these cycles is extremely complex, but there is widespread agreement that they define the series of ice ages over the last 500 000 years, with four ice ages each lasting about 100 000 years, separated by interglacials of some 10 000 years.

The last ice age peaked 18 000 years ago, with ice grinding down south of the Great Lakes in North America and to a little north of London in Europe. Sea levels were again some 300 ft (90 m) below current levels, creating land bridges between North America and Asia and between New Guinea and Australia. We are currently approaching the end of an interglacial period that began about 10 000 years ago.

rose, and after repeated inundations that ended only some 4 million years ago, the Tethys retreated into isolated seas – today's Mediterranean, Black Sea, Caspian and Aral. When the collision between the African-Indian plate and the European-Asian plate occurred, some 20-10 million years ago, great new ranges arose – the Alps, and the whole Tibetan massif of the Himalayas, the Pamirs and the Hindu Kush.

In the west, North America was still pushing against the Pacific plate, creating a bow-wave of instability down the continent's west coast. From Mexico to Alaska, earthquakes and volcanoes punctuated the creation and re-creation of new mountains, which stretched a third of the way across the continent. Because the two plates were moving both over and past each other, the details are hugely complex. But the results are all too evident, in California's earthquakes, in eruptions like

LIFE 5 ITSELF

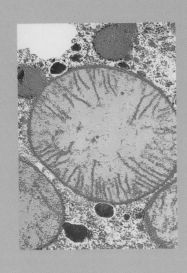

MITOCHONDRIA *These micro-organisms use oxygen to release energy.*

AS SOON AS LIFE EMBARKED ON THE LONG PROCESS OF EVOLUTION, IT REVEALED AN ASTONISHING TALENT FOR MODIFYING AND ENRICHING ITS ENVIRONMENT, AND RECYCLING ITS BASIC REQUIREMENTS, THUS INCREASING THE POSSIBILITIES FOR EVER MORE COMPLEX LIFE FORMS. IT TOOK 3 BILLION YEARS FOR THE CREATION OF THE RIGHT CONDITIONS, BUT ONCE THESE WERE IN PLACE EVOLUTION SWITCHED UP A GEAR, WITH MAJOR CHANGES TAKING PLACE OVER HUNDREDS OF MILLIONS, RATHER THAN THOUSANDS OF MILLIONS, OF YEARS. WITH THE EMERGENCE OF HUMAN BEINGS, CHANGES ARE NOW MEASURED IN DECADES, WITH CONSEQUENCES THAT CAN ONLY BE GUESSED AT.

FOSSIL CREATURE *Hallucigenia lived over 500 million years ago.*

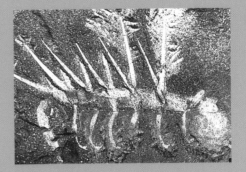

THE LIVING EARTH

Earth's earliest atmosphere was not congenial to life.

But when single-celled life forms emerged in the oceans,

they began to reprocess the sea and air in a way that

allowed their successor species to colonise the land.

Wherever life came from, in the Earth it found a planet that was just right. The basic conditions for life already existed: a second or third-generation Solar System containing a rich mix of elements; a star of such a size that it would burn steadily for billions of years; a planet that had formed far enough from the Sun to avoid being burned, and close enough not to be frozen; and one that had a circular orbit, so that it did not have too great a range of temperatures – it would be hard to imagine life evolving on Earth if half its year was icy and the other half boiling hot.

In this welcoming environment life took root, and became an integral part of the Earth's history. Though the details of those distant times can never be known, since so little hard evidence remains, it is possible to play detective, working backwards from what happened later. In the past three decades, most scientists – notably Professor James Lovelock – have come to see that life played a vital role in its own evolution.

Though not all would follow Lovelock in treating the Earth as a living entity – Gaia, as he called her – none now would deny life's role in modifying the Earth's chemistry to its own advantage. The story that follows summarises the way evolving life, represented by microscopic forms, renewed the atmosphere and contributed to the strata and sediments beneath the surface. In a number of other ways as well, life and Earth's processes are interrelated. Though the actual chain of cause and effect is often little more than speculation, few would dispute that life has controlled the amounts of several major elements and minerals, and modified temperature and climate. Gaia is not at the mercy of the Universe and her own body chemistry. She looks after her own health, and life is one of her tools.

The relationship between life and non-life dates from the Earth's extreme youth. If, some 3.5 billion years ago, the Earth had been lifeless, it would have had an atmosphere rich in carbon dioxide blasted out by volcanoes. And because carbon dioxide is an effective greenhouse gas, the Earth should have increased steadily in temperature, but it did not. How did this come about?

THE EARLIEST LIFE FORMS

The minute forms of life that emerged about 3.6 billion years ago consisted of two types of bacteria: decomposers, which consumed organic matter and excreted methane and carbon dioxide; and photosynthesisers, which used carbon dioxide

FIRST LIFE *Among the earliest life forms are micro-fossils $1/100$-$2/100$ mm across, found in Canadian flints some 2 billion years old.*

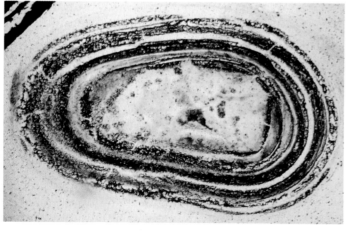

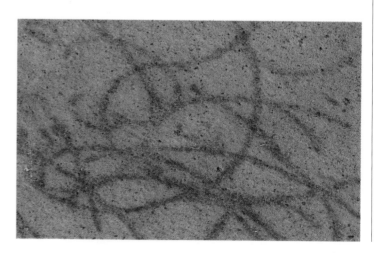

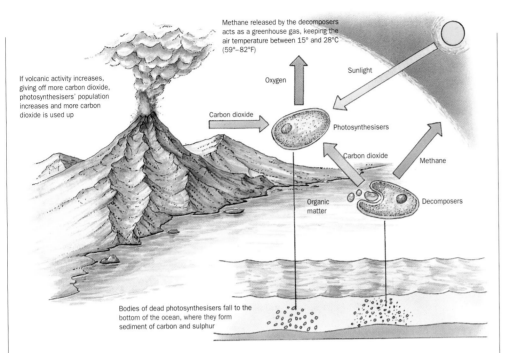

If volcanic activity increases, giving off more carbon dioxide, photosynthesisers' population increases and more carbon dioxide is used up

Methane released by the decomposers acts as a greenhouse gas, keeping the air temperature between 15° and 28°C (59°–82°F)

Sunlight

Oxygen

Carbon dioxide

Photosynthesisers

Carbon dioxide

Methane

Organic matter

Decomposers

Bodies of dead photosynthesisers fall to the bottom of the ocean, where they form sediment of carbon and sulphur

STATE OF BALANCE *While the decomposers gave off carbon dioxide and methane, the photosynthesisers used up carbon dioxide, keeping the Earth's temperature steady.*

and sunlight and gave off oxygen. The interaction between these life forms would quite quickly have created the Earth's third atmosphere, one with less carbon dioxide and increased methane. The sudden decline in the amount of carbon dioxide allowed heat to escape, cooling the planet. The increase in methane, which is also a greenhouse gas, established a new balance.

In brief, even these primitive life forms, which could live only in a critical temperature range of about 15°–28°C (59°–82°F), created their own environment by establishing a positive feedback. For instance, an outburst of vulcanism (volcanic activity) would produce a sudden increase in carbon dioxide. This both raised the temperature on Earth and provided more food for the photosynthesisers. The latter would have increased in numbers in relation to the decomposers, consumed a greater amount of carbon dioxide, and thus brought the temperature back into equilibrium.

Had alien astronomers observed the Earth they would have seen that the atmosphere was in some way not natural to a planet that was merely a geological test tube. They would have known that the Earth was already alive, and well able to look after itself. For another 1.1 billion years – almost a quarter of the Earth's life history – bacteria and algae kept their world in biochemical

balance, and did so despite the continuing bombardment from space by the detritus left over from the formation of the Solar System. If there had been only one major impact every 100 million years – and almost certainly there would have been more – life must have been nearly obliterated many times. Yet the balance was maintained.

The presence of life may have been vital in another way. If the Earth had been left lifeless, volcanoes bubbling up from beneath the seas would have initiated reactions that would have sent hydrogen to the surface, where, in the absence of other gases to combine with, it would have been free. As the fate of Venus shows, if hydrogen is allowed to float free, it escapes into space, making it impossible for any oxygen to combine with it to form water. Earth might well have become as hellish as Venus. In fact, the presence of photosynthesising bacteria saved it from this fate. The process of photosynthesis split carbon dioxide into carbon and oxygen. Oxygen captures hydrogen and makes water, thus ensuring the existence of the oceans.

THE COMING OF OXYGEN

The next stage was the great switchover to a modern atmosphere – a process that involved a decline in methane and an increase in oxygen, establishing a reservoir

of the latter vastly in excess of anything needed to preserve the oceans. Previously, the only source of free oxygen was when it became separated by the breakdown of carbon dioxide and sulphur. Yet these compounds do not split up easily, so most of the oxygen remained bound up until the coming of life. Photosynthesis split the compounds and released the oxygen, while the other elements were absorbed by the photosynthesising bacteria and remained bound up when the bacteria died and drifted to the floor of the ocean. A new process had begun: the release of oxygen and the burial of carbon. On the assumption that life, though microscopic, was as active then as it is now, this process is estimated to have buried about 100 million tons of carbon a year, and released some 260 million tons of free oxygen.

Originally a poisonous by-product excreted by certain types of bacteria, oxygen then became a source of energy for those life forms that consumed the photosynthesisers. The consumers invented a new body chemistry that utilised oxygen, and at the same time evolved far greater complexity – cells with a nucleus, the so-called eucaryotic cells. The change marked a boundary between two different biochemistries, the earlier one belonging to the Archean time (some 3.8 to 2.5 billion years ago), the later to the Proterozoic, which followed it.

The oxygen-consumers achieved their new biochemistry by hijacking simpler cells and incorporating them into what was, in effect, a micro-community. One unit in these communities was the mitochondrian, a bacterium that was once perhaps nothing more than an infection – until its talents proved handy to its host. Mitochondria use oxygen to burn sugars more efficiently than does any other unit of life, so they were incorporated by eucaryotic cells, along with the

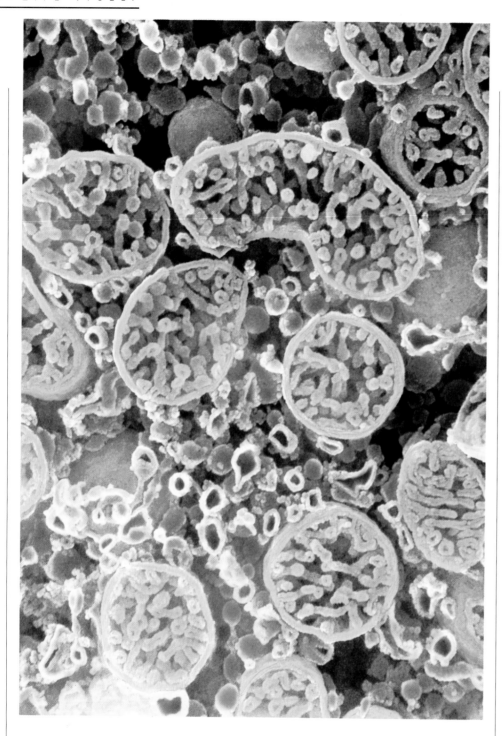

HELPFUL PARASITES
Mitochondria developed from bacteria that invaded single cells, and then evolved to play essential roles in their host-cells.

cells' own genetic instructions. This step was momentous, for it demanded the coordination of complex genetic instructions and a new way of dividing those instructions – the basis for all later and higher life forms, and the basis for sexual reproduction.

The major significance of sexual reproduction is that it makes possible a vast number of genetic combinations. For example, the occurrence of ten mutations in a population of asexual organisms – those that preceded the eucaryotes – could result in no more than 11 types, the original and the ten mutants. But ten mutations in a sexual population could combine to produce 60 000 genetically distinct types. The speed and efficiency of evolution could shift into high gear.

This new beginning did not spell an end for the Archean bacteria – the decomposers. They lived on wherever they could escape oxygen – and still do, in the ecosystems that dominate the biochemistries of the sea floor, of marshlands, and of the guts of most higher forms of animals, human beings included. In fact, any alien astronomer observing the Earth now would know they are here, because the methane they produce is still here. If they were not still here, all the methane in the atmosphere would have been absorbed by the oxygen long ago.

The changeover from methane to oxygen in the atmosphere occurred so long ago, and left so few traces, that no one claims to be able to pin it down even to the nearest 100 million years. But scientists agree that it did occur, and that it was well under way by about 900 million years ago. They have devised a scenario to explain what happened, and why.

Oxygen came to dominate methane, and may well have done so quite quickly because the Earth fell into a prolonged ice age around 2.3 billion years ago. This could have been kick-started by a drop in the level of methane and the consequent cooling brought on by the decline in this greenhouse gas. In any event, the oxygen produced by the bacteria finally overwhelmed the capacity of the oceanic minerals to absorb it. Oxygen accumulated quite quickly, until it became a significant part of the air. The increase in oxygen forced further change. Methane-based cells (the decomposers) declined. Photosynthesisers and those forms that consumed them boomed. The carbon dioxide level fell slightly while the oxygen level rose, and then – because oxygen becomes toxic in high doses – stabilised at something well below its current level (such tiny organisms did not need much oxygen to power their biochemical reactions).

The fact that this process occurred received additional proof from a bizarre discovery during the 1970s. In Gabon, near the town of Mounana, in plains through which the Ogooue river flows from lowlying, forested hills, the French had opened up a uranium mine, named Oklo. When analysed, the uranium was found to be strangely deficient in the form that is most fissionable, and thus the most vital for the nuclear industry – U^{235}. 'Imagine the shock,' writes Lovelock in his assessment of the discovery. 'Had some clandestine group in Africa or France found a way to extract the potent fissionable isotope, and were they now storing this for use in terrorist nuclear weapons?' The answer was no.

In some circumstances, uranium is water-soluble. When these circumstances arose, the uranium began to dissolve and flowed in a stream into a mat of local bacteria, which absorbed it, along with other elements, thus concentrating it. When it reached a critical level of concentration, it began to interact with itself, producing heat. The reaction was limited by the flow of water. The area became a natural water-moderated nuclear reactor, producing heat for millions of years.

Now to the main point: U^{235} is soluble in water *only in the presence of oxygen.* If U^{235} could dissolve without oxygen present, the bacteria might have made use of it much sooner, when there was a good deal more of the highly fissionable U^{235} about. The

EARLY COLONIES *Stromatolites, the first examples of communities formed by cells, still exist today. This one is in Shark Bay, Western Australia.*

result would have been quite extraordinary – a naturally produced atomic explosion. As it was, by the time the uranium became available to the bacteria, it had decayed to a safer level. And by the time the operation shut down, it had used up a good deal of the U^{235}, leaving enduring proof of the arrival of oxygen in Earth's early atmosphere.

On a later visit, the alien astronomers would have found a very different world from the preoxygen version: blue skies, a blue-grey sea fringed by shallows studded with shelly, coral-like reefs of stromatolites (limestone structures). The land would have been speckled with green, yellow and brown mats of algae. The Sun had increased its output to its present level, and microorganisms on the land would have been protected from the worst effects of solar radiation by the newly formed ozone layer.

Through all of this immense period – 1 billion years – there would have been scattered cataclysms caused by the impact of asteroids. But even major impacts – initiating ice ages, cracking open the crust, splashing rocks back out into space, bringing a rain of debris and ash over the whole Earth – were not enough to upset the processes that life had set in place.

MODIFYING THE OCEANS

At the same time, many other processes by which early life forms concentrated elements to form minerals were at work in the oceans. One example is the formation of calcium, an essential component in bones and teeth, and also important in many physiological processes, such as cell division and blood-clotting. Yet it is also toxic – a concentration of calcium ions over just a few parts per million, far less than exists in the oceans, is lethal. For this reason, early life forms evolved the ability to excrete calcium as calcium carbonate, the stuff of seashells – and limestone. The result of this

SAFE REFUGE *Marshlands provide an environment for methane-producing bacteria where they can escape Earth's oxygen-rich atmosphere.*

activity was to create whole limestone 'cities' that colonised shallows around the world. The laying down of limestone reefs, however, may also have been associated with other processes – salt production being one.

Like many minerals, salt is both essential to life and poisonous to it in excess. Despite anecdotal evidence that some rare

CORAL CITIES *Flower-like corals (right) absorb carbon from the water and secrete external skeletons, via which they link up to build vast carbonate rock walls such as the Great Barrier Reef (below).*

people can tolerate drinking seawater in small but life-preserving quantities, on the whole as a species we cannot. Nor can other living things. Few organisms can metabolise salt if it makes up more than about 6 per cent of their weight, whether or not they live in salt water. Most fish live either in fresh water or in salt water, but few can manage both. Too much salt in a cell dehydrates it to the point of collapse; too little salt allows in too much water, until it reaches bursting point. Management of the salt budget is a tricky operation, and variation in salinity is hard to handle.

Since salts are washed down from the land to the sea, and since evaporation leaves the salts there, the sea should, in theory, become increasingly salty, like the Dead Sea or the Great Salt Lake, making survival there steadily tougher. However, this did not happen. Salinity has remained remarkably stable over millions of years.

One reason for this is that all seas have shallow backwaters from which water evaporates, leaving salt deposits behind – the source of today's packaged table salt. This is a self-regulating process. If sea levels fall – as they do during ice ages – you would expect the saltiness of the oceans to increase. In fact, more backwaters are exposed, and more areas are opened up for salt to be deposited, forming sediments that lock away the salt in rocks.

However, this is not the whole story. One form of shell is created by coral. Coral reefs, built by small, soft-bodied polyps, grow near the surface of warm oceans, at depths of about 15-150 ft (4-45 m). They form very effective underwater constructions (witness the Great Barrier Reef and any number of coral islands). Though living reefs cannot project above the sea, they emerge when the

ANCIENT ENERGY SOURCE

Coal is a form of energy rich in carbon, which is captured by plants from the atmosphere, the Earth and sunlight. It was formed from plant material that decayed in swamps or under water, which excluded oxygen from the process and prevented total decomposition. Coal exists in many forms, depending on the type of plant and the conditions in which it fossilised.

The first great coal beds were laid down during a time of warmth and humidity between about 315 and 290 million years ago, when the landmass of Gondwanaland straddling the tropics was partially covered with steamy swathes of newly evolved ferns. These fern forests stretched across areas that now form parts of European Russia, Poland, Britain and eastern North America.

As they died, the ferns matted into peat, which was then flooded repeatedly by the sea and overlaid by layers of sediment. The increasing weight compressed the peat, first into low-grade coal known as lignite, then into ever-harder forms with ever-higher densities of carbon, from ordinary bituminous coal to anthracite.

Although other types of coal formed at other times and in other places, it was the discovery in 1822 of the tropical coal beds beneath England and Wales that inspired the name given to this period, the Carboniferous, which now extends back in time to some 350 million years ago. In Britain, these seams of stored energy had already become the fuel of the Industrial Revolution. Their equivalents in the rest of Europe and America inspired

FOSSILISED SWAMPS *Before forming coal, the great swamp forests of the Carboniferous rotted into peat, as in this quarry in the Falklands.*

phenomenal growth, and even after the burning of billions of tons, there are no signs that supplies are nearing exhaustion – though the release of vast amounts of carbon into the atmosphere could have irreversible effects on the Earth's climate.

sea level falls, creating lagoons in which sediments gather. The isolated backwaters gradually become evaporation pans where salt is deposited. Thus, possibly from Palaeozoic times (550-245 million years ago), living creatures have helped to engineer the conditions that have controlled the levels of the most damaging salts in the oceans.

It is a remarkable idea – that communities of tiny creatures should have been responsible for major feats of planetary engineering. Life and geology became one. And, for unrelated reasons, they remain so, for by coincidence the same polyps that formed the coral reefs and created evaporation pans have also been responsible for

making products that humans find extremely useful – salt and fertiliser.

Once cells were able to live together in communities, they could specialise and evolve larger communities. They could put down roots and grow upwards. In the sea, they could form free-floating cell-cities, with outlying colonies reaching into the depths. In both cases, the cell communities could colonise the third dimension. In other words, they evolved into plants and the first animals.

The relationship between the oxygen-producers (the plants) and the oxygen-consumers (the plant-eaters) now became more complex. Plants produced more

FOSSILS IN COAL *Coal seams sometimes retain the delicate imprints of some of the plants from which they were formed.*

oxygen, and the new forms of life adapted to make use of it. The oxygen-consumers could perhaps have done with even more; as athletes and women in labour know, muscles can make use of a good deal more oxygen than they can get from current levels in the air. But increases in oxygen soon cause it to reach danger levels. Oxygen in large amounts – over 25 per cent of the atmosphere – is flammable; over about 40 per cent it is also toxic. Plants cannot afford to be *too* successful or they will raise the oxygen level to the point at which they risk perishing in a universal blaze; or if they avoid that fate, poisoning themselves. Despite evolving numerous strategies to avoid being eaten, plants need the oxygen-consumers to limit the oxygen level and keep the carbon dioxide level topped up. Between the two opposing forces, a balance was struck.

The new balance was reached when oxygen levelled off at 21 per cent of the atmosphere, where it has remained for the last 800 million years. It was a fair compromise, for it allowed forests to burn sometimes, as

ANIMAL GRAVEYARDS *Limestone is formed from the buildup of shells on the seabed. Both shells and fossils are found embedded in it.*

geologists know from the discovery of ancient beds of charcoal. Burning became possible when the oxygen content reached 15 per cent, but the oxygen level stopped short of the 25 per cent mark and the risk of total destruction by fire. Several tree species evolved to make use of fire. Some conifers and eucalyptuses produce resin-rich materials that are easily ignited by lightning strikes. The fires they encourage destroy small, competing species, but leave the tall trees undamaged. Moreover, some conifer seeds need the heat generated by fires to split them open so that they can germinate.

CARBON DIOXIDE

While the oxygen content of the atmosphere increased, carbon dioxide, which had once made up almost all Earth's gases, underwent a sharp decline. Venus has 300 000 times Earth's current carbon dioxide level, and the early Earth had 1000 times as much, all of it derived from volcanoes. On an Earth without life, weathering would have exposed rocks, which in turn would have reacted with the carbon dioxide to reduce it – but not enough to explain current levels, which are just 0.3 per cent of the atmosphere.

What actually happened was that living things absorbed the gas and carried it into the ground. The process established by life over 2 billion years ago is still working today. Soil everywhere is between 10 and 40 times richer in carbon dioxide than is the air. Once in the ground, the carbon dioxide reacts with calcium silicate (a form of calcium) to form calcium carbonate and silicic acid. In this new mineral state it is washed down to the sea, where sea creatures by the billion turn it into their shells.

In death, the animal remains rain down to the seabed, where they form limestone.

This is a highly efficient process, so efficient that when the Earth's plant cover is at a maximum, the carbon dioxide level drops, depriving the plants of one element in the chemical recipe needed for photosynthesis. To compensate for this, some 10 million years ago, a whole new category of plants emerged, including many of today's grasses, which can photosynthesise at lower levels of carbon dioxide.

Low levels of carbon dioxide in the atmosphere therefore correspond with high levels of plant cover. Less carbon dioxide means less of a warming blanket round the Earth, which means that it becomes cooler. The unlikely conclusion to this argument is that plants thrive on ice ages, which seems to fly in the face of common sense, because during ice ages the spreading ice sheets would seem to deprive the plants of space to live. In fact, the ice sheets suck up so much water that they expose vast expanses of continental shelf – equivalent to an area the size of Africa. Moreover, over the past few hundred million years, the exposed area was in the tropics. Paradoxically, ice ages coincided with the growth of tropical forests, and with evolutionary pressures that produced yet more plants.

THE SULPHUR CYCLE

Among the many processes by which life interacts with and regulates non-life is one that controls the circulation of sulphur. Sulphur is essential to plants and animals. Protein consists of some 20 amino acids, two of which contain sulphur. It is also a constituent of thiamin (vitamin B_1) and various enzymes, the substances involved in many important biochemical reactions. Animals obtain sulphur from plants. Plants get it from the soil – each acre of farm crops

needs about 26 lb (12 kg) of sulphur. The soil gets it in part from volcanoes and the breakdown of minerals (and in part, now, from artificial fertilisers). But rain washes sulphur away about three times faster than the soil can release it. So how does the sulphur get back into the land, other than in a farmer's truck?

IMPOSSIBLE ODDS

According to the British astronomer Professor Fred Hoyle the chance of life forming through an accidental shuffling around of molecules is about as likely as a whirlwind in an aircraft factory blowing together the components of a working jumbo jet.

One answer is that sulphur, in various chemical forms, drains into marshes and estuaries, where complex reactions occur that allow it to be absorbed by water and returned to the land in rain. But this cannot

RICH GRASSLANDS *Grasses, like those of the Ngorongoro Crater, Tanzania, have evolved to survive on low levels of carbon dioxide.*

be a complete explanation because a major derivative of sulphur is hydrogen sulphide, which is poisonous, revoltingly smelly and explosive. Some swamps are indeed smelly – hydrogen sulphide is usually described as smelling like rotten eggs – but they do not poison you and they do not explode.

It turns out that hydrogen sulphide is produced at sea. Basing their research on a type of red seaweed called *Polysiphonia* and several types of single-celled, free-floating plants, scientists theorise that the sequence started with shallow-water plants that found themselves stranded at low tides. As they dried out, desiccation concentrated their salt content. The plants' response was to evolve the use of chemicals that prevented desiccation. One of these chemicals, when released into seawater, reacts to produce a form of sulphur. When it is drawn up by

LIMESTONE: THE FOUNDATION STONE

One way in which life has modified Earth's chemistry is by the control of calcium. Calcium occurs in several forms. One of these, calcium carbonate, is extracted from water by plants and animals, which use it to make their shells or other hard parts. The rain of shelly material onto the ocean floors when the organisms die has created immense sheets of limestone in many forms, including chalk, a soft type of limestone, effectively removing calcium from the evolving Earth.

This process can only occur in warm, shallow seas, however, such

CARBON STORE *The limestone pavement at Malham, Yorkshire, was formed under a tropical sea by a process that removed carbon from water.*

as those that covered continental areas in the Carboniferous (350 to 315 million years ago) and in the late Cretaceous between 80 and 65 million years ago.

When the seas receded and land was pushed up, the limestone was thrust to the surface, emerging in places in stark limestone pavements such as the Burren in County Clare, western Ireland, one of many natural wonders created by limestone. Here, 500 sq miles (1300 km²) of the Irish 'basement' of limestone was scraped clean by glaciers some 15 000 years ago, forming a bare, cracked moonscape. Since water quickly percolates through the rock, only specialised plants such as foxgloves and rock roses can grow there. Beneath, the water has eroded a maze of caves. A similar process created the immense

chalk sediments that are such a feature of northern Europe, such as the White Cliffs of Dover and the Downs in south-east England.

Limestone and its derivatives have always had a wide range of practical uses, particularly in building and, in recent times, in industrial processes. The pyramids of Giza in Egypt are made of limestone blocks, and the stones of Westminster Abbey and Buckingham Palace are limestone, from quarries in Caen, France. Countless modern buildings are faced with it. Crushed limestone is used to make the foundations for roads, railways,

BIG BRICKS *The pyramids of Giza in Egypt were built from great slabs of limestone averaging 2½ tons in weight each and quarried nearby.*

dams and airport runways. Lime, produced by heating limestone, is crucial in the manufacture of cement, aluminium and glass, and in animal feeds and fertilisers. And without lime putty, the Italian Renaissance masters could not have made their frescoes.

evaporation, a further chemical reaction produces microscopic sulphuric particles, which act as 'seeds' for the condensation of water droplets. Hey presto – enough sulphur-rich rain is produced to balance the Earth's sulphur budget.

THE IODINE CYCLE

Another essential element is iodine. Ever since the first amphibians appeared, this mineral has been vital to development. It regulates the development of tadpole into frog – deprive the tadpole of the iodine-rich hormone, thyroxin, and it will not metamorphose. In humans, iodine is used to manufacture thyroxin in the thyroid gland. A deficiency lowers the metabolic rate, causes goitre and can lead to a condition known as cretinism – all this for want of 1.7 *thousandth* of an ounce of iodine.

The richest sources of iodine are some types of sea creature – cod, oysters, halibut, salmon – sea plants such as seaweed, and salt. The sea itself is relatively rich in iodine, containing it in a proportion of 1:8000, which is a good deal more than exists in the soil and rocks. This is not surprising, perhaps, because for 3 billion years, rivers have been washing iodine into the sea. If that were the end of the story, all the iodine would be in the sea by now, and all of us land creatures would have gross thyroid deficiencies. Once again, life processes developed that ensured the recycling of iodine. The crucial player is seaweed. Seaweeds are a type of giant algae, and are among the simplest forms of plant life. Iodine in seaweeds is metabolised and released as a gas, methyl iodide, into the atmosphere. Some of this is incorporated into rain and returned to the land.

These are just some of the interrelated traits that the Earth developed as life evolved, and most of them were in existence before the emergence of the earliest forms of life as we know it: marine invertebrate creatures that were not microscopic, and which emerged at the start of the Cambrian Era, some 550 million years ago (although actual dates are much disputed, estimates for the start of the Cambrian vary from 570 to 535 million years ago).

MINERAL RECYCLERS
Polysiphonia (right) releases sulphur into the sea. The intertidal zone (below) is colonised by seaweeds that draw iodine from the sea and release it into the air.

THE PATTERNS OF EVOLUTION

The increasing chemical complexity of the Earth provided

an ever-increasing number of different niches for plants

and animals. Marine creatures extended their range, plants

colonised land, and animals followed, in ever greater variety.

Roughly 550 million years ago, at the start of the Palaeozoic ('Ancient Life') – once, but no longer, seen as the starting point of life itself – the main landmass was the unified block of Gondwanaland, linking Africa, South America, India, Antarctica and Australia. Northern Europe, Siberia and North America were separate islands. All these landmasses were set along the tropics and temperate regions, allowing the free circulation of ocean currents, keeping the poles free of ice, and ensuring high sea levels. Nurtured by these extensive, warm, shallow seas, marine life exploded. Creatures developed the ability

to build shells, giving them protection and providing a firm anchor for muscles, which in turn enabled them to move better. For the first time, life formed fossil records in sedimentary rocks – and in doing so used up the lime (calcium carbonate) in the seas, and changed the mineral content of the early oceans.

Fossils are not, of course, the creatures themselves. For one thing, the soft parts almost always rotted away quickly. For

another, the hard parts are not preserved as bone, but as rock. They are formed when minerals in the surrounding ocean, sand or mud react with, permeate and eventually replace the original shell or bone. What is left is something that is not the original, but a record of the original in rock.

Fossils of early sea creatures were first identified in the early part of the 19th century in Wales (*Cambria* in Latin, which is why the period they are associated with, the first part of the Palaeozoic, was named the Cambrian). As it turned out, 'Cambria' was not the best source of Cambrian fossils, which survived particularly well in what is now British Columbia.

Like all continental cores, the North American shield – consisting of Canada and Greenland – rode clear of the surrounding ocean, while the more southerly areas, the land that would one day form parts of Canada and the United States, formed the continental shelf. Here, fine-grained mud at the bottom of a 20 mile (32 km) long underwater cliff created smooth sediments that proved an excellent source of fossils which were first discovered in the early 20th century. These rocks, known as the Burgess

ANCESTRAL PRAWN *One of the commonest fossils of Cambrian times was Canadaspis, a 3 in (7.5 cm) remote relative of shrimps, prawns and lobsters.*

Shale, record a wonderful range of small marine creatures, the ancestors of all modern invertebrates.

Some, the brachiopods, looked like small oysters and clams, while others – trilobites – were louse-like, with hard external skeletons, and scavenged the ocean floor, grazing on algae and seaweed. Trilobites, which make up about 30 per cent of all Cambrian fossils, had a three-lobed shell (as their name suggests), eyes and a spiny tail. The design worked well, and they evolved a huge range of species, some acquiring a tail-paddle obviously adapted for swimming rather than crawling. Though the trilobites themselves later became extinct, they have their descendants among land-based creatures – the modern insects.

No fish were found in the Burgess Shale, but one of the creatures there is the oldest known animal to have a primitive spinal column, something that would prove among the most successful of evolutionary developments. Some 450 million years ago, these early chordates – our ancestors – did not look a promising bunch. They were tiny, boneless things with only a slight hardening to indicate their new anatomical invention.

Within some 50 million years, however, the rocks record the arrival of proper, if primitive, fish – jawless ancestors of lampreys with bony armour. They proved the advantages of a backbone: its stiff, board-like design provided a firm base for muscles, allowing a sinuous motion ideal for swimming. For a while, their slow and solid ancestry doomed them to grow bony armour

SEA-FLOOR SWEEPERS
Louse-like trilobites such as these Elrathia found in Utah were common marine scavengers 500 million years ago.

as a defence against predators – the Silurian seas were home to fearsome 10 ft (3 m) long sea scorpions, the eurypterids – but with the passage of another 50 million years, through the Devonian period (which ended around 360 million years ago), fish discovered that fleetness of fin was a better defence than armour plating. Vestiges of that armour plating still survive, however, in the scales and bony head regions of modern fish.

SEA LILIES *Crinoids, with their stalks and five arms, were remote Palaeozoic ancestors of today's echinoderms such as feather stars and starfish.*

The Devonian has been called the Age of Fishes, not because there were so many species (there are far more today), but because they were the dominant life form. Ancient lineages such as the jawless fish were joined by fish with jaws, ancestral sharks (making them one of the most ancient of vertebrate animals), and fish with proper

THE BURGESS SHALE

One of the greatest fossil treasure sites is the Burgess Shale, in the Yoho National Park in British Columbia. Here life, and its sudden diversification, is recorded from just after the appearance of the first multicellular animals some 570 million years ago. The fossils are unique in that they preserve both the hard and soft parts of animals, and even their stomach contents.

The site, a quarry, lies 8000 ft (2500 m) up in the Rockies and was discovered in 1909 by Charles Doolittle Walcott, head of the Smithsonian Institution. Walcott named the quarry after a nearby pass recalling a 19th-century governor general and the fossils, such as Wiwaxia and Takakkawia, after local peaks and lakes.

CAMBRIAN IMPRINTS *Wiwaxia (below), which crawled along the sea floor, was covered with scales and had two rows of spines along its back. Canadia (below right), a bristly worm, crawled and swam.*

The shallow-water creatures, most only a few inches long, lived on mudbanks at the base of an almost sheer reef. When they died, they were swept away and buried by mudslides, depriving the creatures' remains of oxygen and compressing them into flat films of carbon, preserving them with microscopic precision. It was not until the 1970s, though, that their true significance was revealed.

All four major kinds of arthropods are here: extinct trilobites; crustacea such as lobsters, crabs and shrimps; spiders and scorpions; and insects. But there are also a score of creatures that do not belong to these groups, including an animal that looked like a flower and another that walked on seven pairs of spines, named *Hallucigenia* for its 'bizarre and dream-like appearance'.

Perhaps the oddest is *Opabinia*, a sort of shrimp-like creature, some 3 in (7.5 cm) long, with a single tentacle, or nozzle. In 1975, modern techniques of dissection and photography revealed that, unlike other Burgess animals, *Opabinia* was no arthropod. This science-fiction animal had five eyes and a nozzle that seems to have acted as a vacuum-cleaner mouth. Its ancestors and its descendants (if any) remain unknown.

The diversity preserved in the shales suggests that early life experimented with five times as many designs as those that survived. In the words of the evolutionist Stephen Jay Gould, this is 'a disparity in anatomical design far exceeding the modern range throughout the world!' Most of the designs died out, and since then, no new designs have emerged. This discovery has contributed to a continuing scientific debate about evolution – the nature of 'progress', the growth of complexity, the sudden bursts of extinctions, and the randomness of survival.

BURIED TREASURE *The Burgess Shale quarry has revealed very early fossils that are unique in their diversity and state of preservation.*

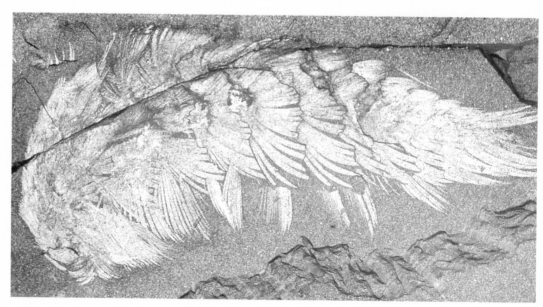

LAND INVADER *Cacops was an early amphibian that exploited the new world that opened up when plants colonised the land some 280 million years ago.*

bones, together making the four groups into which all fish today are divided.

An alien scuba-diver in the Devonian would have seen giant reefs of archaic coral growing in warm, clear waters, and crinoids, or sea lilies, ancestors of starfish and sea urchins, looking like huge flowers on long stems. Cruising among the crinoids were nautiloids with scrolled shells searching for trilobites. Everywhere were the newly evolved shelly ammonites, or free-floating molluscs. Occasionally the dark shapes of sharks, very similar to today's, would have loomed up, or even monstrous 50 ft (15 m) long carnivorous placoderms, an extinct class of fish characterised by bony armour plates on the head and body. And darting everywhere were scores of fish species, most of which would look quite familiar to inexpert modern eyes. Some – the ancestors of modern lungfish – had already begun to develop lungs as a way of drawing oxygen directly from the air. And perhaps, here and there, the diver would spot a curious-looking fish with lobed fins which, with the experience of hindsight, he would recognise as the precursors of legs.

At the same time, the land, too, acquired a new set of colonists. In addition to the simple rock-hugging algae, the first upright plants appeared, exploring the third dimension with roots and branches. Giant fern-like trees of horsetails and club mosses formed the first forests. In the beginning, they were dependent on water as a means of reproduction. But by the late Devonian, some ferns had evolved hard seeds that grew independently of water. Among the forest roots scurried the first land animals of which fossils survive – scorpion-like creatures, perhaps scavenging for worms and small crustacea.

Finally, in the course of some 100 million years, as the scattered island-continents of the world approached each other, as the great sweep of the Caledonian mountains rose to form present-day Scotland and fringe Greenland, Scandinavia and North America's east coast, fish evolved with lungs and with muscular fins that they could use for walking. Some 350 million years ago, a new type of animal emerged, still rooted in water but with lungs and legs: the first amphibians had appeared.

So successful were these creatures that they spread rapidly, spawning new species to fill the niches offered by this new way of life. All amphibians, though, need a moist habitat and water to breed in. With skins unprotected by scales or hair, and producing soft, gelatinous eggs, they and their offspring could not cope with extremes. Prone to freezing or dehydration, the earliest land colonists could never stray far from the swamps, lakes and coastal shallows from which their ancestors had emerged.

LIVING ON LAND

Two innovations were needed to free land vertebrates from their watery shackles: a tough, drought-resistant skin and an egg capable of developing on dry land. Both these adaptations evolved over the next 50 million years, producing the first reptiles.

Thus liberated from the need to colonise two environments at once – water and land – the reptiles began to spread rapidly. At first, they were probably all carnivores, preying on each other and on amphibians, who offered well-packaged sources of energy. Newly evolved carnivores faced a problem, though. There were only limited numbers of other animals. Under growing evolutionary pressure, some species discovered a cornucopia: the plants of the Carboniferous that between 350 and 290 million years ago colonised most of the land, laying down the coal seams that give the period its name. Here lay a huge evolutionary potential that, once realised, would support countless new species.

As food sources, plants present problems. Some have to remain in the digestive

EARLY CONE-TREE *A fossil cone from a club moss provides evidence that these tree-like ferns were common in Carboniferous forests.*

system a long time while they are broken down by bacteria, and a great deal of vegetation has to be eaten to match the energy derived from meat. Thus herbivores tend to be large and heavy by comparison with carnivores, a tendency that also served as a defence. In response, the carnivores evolved larger forms themselves. Quite rapidly, evolutionary pressures produced a vast range of reptilians.

One of the problems for both reptiles and amphibians is temperature control. Both are cold-blooded, which means that they do not generate much metabolic heat, and certainly not enough to be independent of their surroundings. Such animals cannot live outside warm climates; after a cold night they rely on the sun to warm them up to working temperature. Clearly, any predator that could generate a little extra heat in order to be that much more active first thing in the morning would be able to catch any amount of prey.

By the relentless logic of evolution, such animals arose. These were the mammal-like reptiles, or protomammals, so called because they had a number of mammalian features, such as canine teeth. In some ways, their amphibian origins are clear. One of their early representatives, common in the fossil record, is a solid, lizard-like animal known as Dimetrodon, one of a whole range of species that had a fin, consisting of a fan of skin supported by spines, sticking up from their backs. The purpose of the fin could well have been to act as both a solar panel and a radiator, for any creature with a device that could gather heat more quickly than other species and get rid of it after a chase would have a clear-cut advantage. However, the existence of the fin reveals the continuing need for one, and thus the fact that these creatures were cold-blooded.

Later protomammals came up with a different solution. They were warm-blooded, almost certainly;

DINOSAUR HUNTER *Robert Bakker was one of the first to argue that dinosaurs were warm-blooded and fast-moving, like mammals.*

'almost' because the whole issue of whether these protomammals and their successors, the dinosaurs, were warm-blooded is one of the great scientific controversies of recent decades. Traditionally, of course, dinosaurs were seen as reptiles, and as such were dismissed as cold-blooded. That went for the protomammals as well. However, many scientists now argue that both types must have been warm-blooded because they share certain features that are associated with later warm-blooded animals, such as the presence of a diaphragm and a secondary palate, which allows an animal to breathe and eat simultaneously.

If the early protomammals were indeed warm-blooded, this marked another major change, for warm-blooded animals move much faster than cold-blooded ones, and can do so at any time of the day or night. At a stroke, potential hunting time doubled and the range of prey that could be caught increased, advantages that more than made up for the fact that warm-blooded animals have to eat much more, in order to power their super-active metabolisms, than cold-blooded ones of the same weight.

THE KARROO

One place that contains a large cross-section of protomammal fossils, and therefore a wealth of evidence for this debate, is the Karroo desert, South Africa's high veldt of scrub, grass, rock and sand. Here, where springbok dart away from cars travelling the lonely route from Cape Town to Johannesburg, sandstones and shales deposited on the sea floor up to 400 million years ago have been thrust up in table-top sections standing at odd angles. In this corner of what was once Gondwanaland, Earth's history from before the time when amphibians emerged is written in

SPIKE BACKED *Dimetrodon had spines along its back, and these probably supported skin that enabled the creature to absorb and lose heat.*

stone. The oldest rocks formed as rubble was dumped in a shallow inland sea by icebergs that broke away from polar glaciers. As Gondwanaland shifted, the glaciers retreated. The land rose, released from its heavy burden of ice, and sediments were washed down from the surrounding highlands to fill the great depression.

Impressions of horsetails, mosses and ferns show how the first plants colonised the bare landscape. Then, in higher strata, appears a wonderful range of amphibians and protomammals that must have migrated into the area from warmer regions as the ice retreated. One of the most successful groups were the herbivorous dicynodonts ('two-tuskers'), which sported a pair of downward-pointing tusks – useful for digging

roots, perhaps – emerging from a parrot-like beak. They must have looked like sabre-toothed tortoises. Perhaps the tusks and the beaks also came in handy for defence, for the dicynodonts were preyed upon by formidable carnivores known as gorgonopsids, named for the Gorgon, the ancient Greek creature of such fearsome ugliness that the very sight of it turned its victims to stone.

So prevalent are the fossils of protomammals in the Karroo that it is possible to estimate roughly how many predator species there were compared to their prey. This relationship is at the heart of the debate on the evolution of warm-bloodedness, as one palaeontologist in particular, Robert Bakker, has argued.

Bakker has devoted his life to a combination of hard research, original theorising and popularisation. Bakker argues that cold-blooded predators need less food than hot-blooded ones, and that in any animal community the predator-prey ratio reflects this fundamental fact. If warm-blooded predators need a large quantity of food, a large number of prey animals will support a comparatively small population of predators; as cold-blooded predators need less food, prey animals will support a relatively

HEAVYWEIGHTS *Elephants gather at a water hole in Namibia's Etosha National Park. On the plains of Africa today, herbivores far outweigh carnivores.*

PERMIAN EXTINCTION *Sea scorpions were among the many victims of the great extinction at the end of the Permian, 250 million years ago.*

cent. Along the way, the protomammals had become metabolic dynamos, eating more and processing it quicker.

Once that happened, some 270 million years ago, evolution again took off, as the fossil remains defining the start of this new period, the Permian, show. As species competed to exploit the newly opened niches, herbivores proliferated in size and shape, pursued literally and figuratively by predators. This evolutionary dance, which spanned some 20 million years, saw the modification of size, fleetness, teeth, defences, eyesight and claws. In the Karroo, the sediments reveal a great range of creatures from small burrowers to grazers, to formidable predators.

THE GREAT EXTINCTION

The dance was in full swing until about 250 million years ago, when suddenly the music came to a stop.

The change was first spotted by a 19th-century English geologist, John Phillips, who analysed the species present in the rocks, and came up with the startling information that the numbers dropped dramatically and suddenly. It was not an easy proposition to prove, because so few creatures are ever fossilised. Perhaps the change preserved diversity, but of a different kind of creature – worms, jellyfish,

larger population of predators. The proportion of predators to prey in any area therefore indicates whether the predators are cold-blooded or warm-blooded.

For example, on the plains of central Africa today, if you add up the weight, or biomass, of the herbivores – all the elephants, giraffes, buffaloes and their like – and then the weight of the carnivores, you discover that the carnivores make up only about 1-3 per cent of the herbivores' biomass. But in a cold-blooded community, such as one in which spiders are the top predators, spiders make up some 20 per cent of the biomass. On average, warm-blooded animals eat, relatively speaking, ten times the amount that cold-blooded ones do. Though the details vary depending on the type of animal, Bakker argues

that, as a rule of thumb, this is a good way to distinguish between the consequences of the two types of metabolism.

According to this argument, the later protomammals of the Karroo were warm-blooded. In the early days, when Dimetrodon and its fin-backed relatives ruled the swamps, some 25 per cent of the animals were predators, indicating that they were cold-blooded. Later, the figure drops to no more than 12 per

DEATH IN THE SEAS *Brachiopods and a trilobite (bottom right) were among the marine creatures that did not survive the Permian extinction.*

slugs, insects, all those soft-bodied animals that seldom get fossilised. The answer, though, was worth striving for, because it could provide information about how many species the Earth can hold, and thus about the health of the modern world. By analysing the past correctly, scientists hoped to understand the present and future better.

In the 100 years since Phillips's time, many more fossils have been collected, and it has become possible to assess diversity in more detail. In general, scientists agree that, all things being equal, species tend to diversify, producing ever more specialised forms. But at the end of the Permian all things were not equal. Whatever happened 250 million years ago, the extinction was real, as fossils in several major rock sequences show.

In the Karroo, 100 different types of fossil vanished, suggesting that almost half the area's species suffered near-simultaneous extinction. In a coda lasting some 5 million years, new species evolved to replace them. But then came a universal silence, a great dying. The whole era that geologists call the Palaeozoic, 'Ancient Life', the 300 million year progression from small marine organisms to protomammals, ended. The protomammals became extinct. At sea, some 90 per cent of all invertebrate families vanished, including almost all corals, brachiopods and sponges. The trilobites, for millions of years the oceans' universal scavengers, vanished utterly. Most families of fish, snails and clams disappeared as well.

The evidence from elsewhere is more enigmatic, for in many places where scientists expected to find strata, they found nothing. Across the USA and Europe, the strata recorded the late Permian, and then the early Triassic, when evolution got back into its stride. But there was nothing showing the changeover from one to the other, a gap of between 2 and 5 million years.

What happened? Was there a slow succession of deaths caused by some slow-acting mechanism? Or rapid extinction? The debate merged with a wider controversy that developed in the late 1970s over why, occasionally, the Earth apparently suffered such appalling losses.

In order to explain the Permian extinctions, an accepted scenario developed. According to this, at the end of the Permian, Gondwanaland crashed into the North America-Eurasia landmass, forming Pangaea, with slow-acting but ultimately catastrophic effects. First, the interior of the supercontinent, now far removed from the moderating effects of the sea, became hotter in summer and colder in winter. Heat and drought in combination brought death. Meanwhile, around the peripheries and the polar regions, other related changes occurred. Ice caps grew and the seas receded, draining shallow internal lakes and exposing huge areas of continental shelf. In these arid conditions, the range of species spawned by marshlands and forests slowly died.

This theory, never proven, began to come apart in the late 1970s when Chinese geologists identified successive layers of Permian and Triassic sediments, separated by a thin band of white clay. Excitement mounted when they announced that the layer was rich in certain rare elements, particularly iridium, which, according to models of terrestrial evolution, should be concentrated in the Earth's core. The excitement arose from the fact that just previously, the father-and-son team of Luis and Walter Alvarez had announced the discovery of an iridium-rich layer of clay near Gubbio, Italy, dating from the time when the dinosaurs went extinct at the end of the Cretaceous 65 million years ago. The clay held 30 times as much cosmic material as did any of the sediments above or below. They related the layer and the

FLOATING SHELLS *Ammonites, like these from the Jurassic strata in Robin Hood's Bay, Yorkshire, floated in ancient seas during the Cretaceous era.*

extinctions to the catastrophic impact made by a type of asteroid, a chondrite, in which iridium had a particularly high concentration. With the discovery of a comparable layer in China, geologists briefly foresaw a universal explanation for extinctions. In fact, later research showed the Chinese layer of clay not to be quite so easily datable, and to be void of the crucial elements.

The mystery about the Permian extinction endured. The traditional explanation received another blow in 1988, however, when Polish geologists described an odd chemical change recorded in the shells of brachiopods, shellfish that were common in the Permian. The change relates to the ratio of a form of carbon, $C13$, to normal carbon, $C12$. When organic productivity –

principally in the form of photosynthesis by plankton in the oceans – is high, great amounts of carbon with high ratios of C13 to C12 are absorbed by living tissue and passed through to shells. The Polish geologists identified a spurt of organic activity towards the end of the Permian, followed by a decline so sharp that it suggests that almost all plankton died. The effects would have rippled all the way up the food chain.

In 1990, geologists in South Africa made a similar study of protomammal teeth and found the same steep decline. The date of the decline is still unclear, but it is possible that the two events have a similar cause, and occurred together. If so, the coincidence and the lack of sedimentary evidence suggest a dramatic possibility – that whatever triggered the great Permian extinction happened so quickly that it left a trace too small to survive the passage of 250 million years.

DEATH AND RECOVERY

By now, on the one-day clock of the Earth's history, there is a mere 80 minutes to run until the present day. During that time, the dinosaurs emerged from the ruins of the Permian age and dominated the Earth for the next 150 million years while the first, shrew-like mammals endured, imprisoned in their few limited niches – inside trees and down holes – until the tides of evolution turned to their advantage.

This happened – many scientists now believe – with the arrival of the asteroid identified by Luis and Walter Alvarez. There was no equivalent mystery here. 'Iridium layers' were found elsewhere – southern Spain, Denmark, New Zealand, at over 50 sites in all by the mid-1980s – together with other evidence of a catastrophic impact. In southern Spain, tiny, iridium-rich spheres of rock seemed to have been blasted from Earth into space, from where they fell back, covering a good

DINOSAUR GRAVEYARD
The remains of a brachiosaurus from the Jurassic lie buried in sediments in the Dinosaur National Monument, Utah.

proportion of the surface. If so, they were the splashings of something extremely big. In other layers of a similar date, fine soot particles were found – the evidence, apparently, of immense forest fires. Many scientists agreed on the conclusion: a massive asteroid had struck the Earth. Then, in the late 1980s, some thought they knew where it had struck: the Yucatan Peninsula, where a giant crater lay buried deep under younger sediments.

The evidence now available on the impact of the asteroid that killed off the dinosaurs also provides a scenario for the Permian extinction. The effects would have been dire: an asteroid 6 miles (10 km) across and travelling at 25 000 mph (40 000 km/h), blasting a 180 mile (290 km) crater, would release 10 000 times as much energy as mankind's total nuclear arsenal, causing

PALL OF DEATH *The burning oil fields in Kuwait after the Gulf War gave a glimpse of the type of clouds that would destroy life after a massive asteroid strike.*

CYCLE OF LIFE AND DEATH *A 40-million-year-old fossil fish is a reminder that, even during stable eras, species die out as new ones appear.*

enough acid rain to poison the top 300 ft (90 m) of ocean and a tidal wave 1000 ft (300 m) high across the proto-Atlantic. The detritus, blasted into orbit and falling as fireballs, would ignite forests worldwide, producing a planet-wide pall of dust like that which hung over Kuwait after the Gulf War. The world fell into a long winter.

The effect on animals was harder to establish. Certainly, there was evidence of terrible consequences. In seas that once covered southern Spain, 50 species of minute sea plants, planktonic foraminifera, vanished. Ammonites vanished en masse. But the fossil evidence from larger species was more enigmatic, for the processes of fossilisation can make gradual extinctions look sudden, fast extinctions look slow.

Gradually, a consensus emerged from the controversy. The impact that ended the Permian had been preceded by several million years of slow climatic change – identified by some with an immense series of eruptions in India that eventually covered 6000 sq miles (15 500 km²) with lava up to 1¹/₂ miles (2.5 km) thick. A steady decline in many species was capped by the hideous consequences of the impact. It was a combination that slew all large animals and over half the rest.

The rebirth that followed the Permian extinction produced creatures of such size and variety that they have dominated the human imagination ever since they were first identified in the early 19th century. There has never been a time when dinosaurs have not been admired, with the sort of admiration reserved for monsters that are safely behind bars. The more that palaeontologists discovered about them, the greater the fascination, for the dinosaurs were part of whole ecologies. The dinosaurs themselves spun off sub-forms, as the mammals were later to do, colonising the air and the sea. So successful were they, and so obviously adapted for speed and for social living, that their traditional designation as 'reptiles' has become increasingly anachronistic, as the whole debate about warm-bloodedness shows. They were clearly much more than reptiles.

On land, the flowering plants emerged – evolving into over 80 per cent of all plants – and insects proliferated. A few ancient ammonite species had evaded the Permian extinction, and now they came into their own, evolving over 1000 genera, each with its own web of species by the dozen, their chambered cells enabling them to respond to minute changes in depth, temperature and salinity. They were everywhere along the shallow continental shelves, as anyone knows who has fossil-hunted in Lyme Regis, or Robin Hood's Bay, or any of dozens of exposed sites around the shores of Europe and America.

Eventually, by the late Cretaceous, the plant and animal communities on Earth were as rich as those of today's world. Indeed, the dinosaurs were well on their way towards intelligence, with upright species evolving forepaws that were almost as mobile as hands and a stance that might, with

ANCIENT FROG *During the Oligocene, some 38-27 million years ago, numerous life forms, such as this frog, would have been familiar.*

further time, have evolved into something oddly human. It was a line of development cut off, with much suffering, when Nemesis struck.

EXTINCTION AND GAIA

The great Permian and Cretaceous extinctions raise an intriguing question. It took 150 million years for the dinosaurs to evolve and die, and only then did the mammals have their chance at domination. If there had been no great extinction at the end of the Permian, would evolution have sustained its progress? Would mammals have somehow emerged sooner from their distant forebears, the protomammals? Would there have been humans on Earth 180 million years ago? Perhaps – but then perhaps not. For one thing, evolution involves

a lot of randomness. It is unlikely that the same solutions would have occurred twice.

But as far as the Earth was concerned the extinctions changed nothing. The impact of another massive asteroid, the destruction of the dinosaurs, the emergence of mammals, the break-up of the Pangaea supercontinent, the slow evolution of modern geography and modern animals, and the emergence of a naked ape with a big brain – none of this altered Gaia's chemistry, the rules of which had been made up when she switched to an oxygen-rich atmosphere.

Of course, each minor adjustment for the Earth meant major adjustments for its plants and animals. The last one occurred when the ice finally withdrew from northern Eurasia and America, beginning some 18 000 years ago. Several species had adapted to the pattern of ice ages – rhinos and mammoths roamed Siberia, clad in woolly coats against the cold. When the climate changed, they died out, leaving not only fossilised remains but also actual remains, deep frozen in the tundra. As the ice melted, leaving a surface scraped and washed bare by ice and glacial meltwater, forests recolonised the northern wastes, turning rock and gravel to thin soil, opening the way at what was once the ice's southern edge for the spread of humans.

For the Earth, though, all this was routine. Ice ages came and went, asteroids struck, but the Earth was not easily upset, and her mechanisms soon righted themselves. Whether the atmosphere was processed by plankton or mammals, the end result was the same – until now.

THE FATE OF THE EARTH

The mechanisms that control Earth's past and present may be to some extent predictable, but the emergence of human beings and an industrialised world, has introduced a random element into attempts to foresee the planet's future.

The story of the Earth's creation and evolution is a dynamic one and, like any biography, implies an end. The Earth will change, and at some time it will die. Change will happen in any number of ways, some growing out of the distant past, some dimly apparent, some random. Whatever the nature of the changes that are to come, there is in the distant future an inevitable end.

Currently the Earth is between ice ages, with the clear implication that, barring an unforeseen and fundamental change in the climate, the world will experience another ice age. But when? On this, scientists find it hard to reach agreement, for predictions demand clear trends, and the trends are anything but clear. Statistics, the spread and retreat of glaciers, the migration of animals and plants – all these provide key clues, or mere confusion.

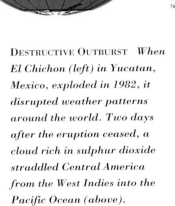

DESTRUCTIVE OUTBURST *When El Chichon (left) in Yucatan, Mexico, exploded in 1982, it disrupted weather patterns around the world. Two days after the eruption ceased, a cloud rich in sulphur dioxide straddled Central America from the West Indies into the Pacific Ocean (above).*

Underlying the conflicting evidence are conditions that remain stable for aeons – the astronomical rhythms apparent in the eccentricities in the Earth's spin identified by Milutin Milankovitch. On this basis, it is clear that the amount of heat falling on the Earth's surface will gradually diminish, until it reaches a low point 23 000 years from now. Refinements of this prediction can be tackled in several different ways. Statistics suggest one line of reasoning: no recent interglacial has lasted more than 12 000 years, which implies that we should be in for only another 2000 ice-free years.

Current trends provide another angle, but a trend is hard to spot. Between 1940 and 1960, the Earth's temperature fell by about 0.1°C, a trend that, if continued, would bring on an ice age in 700 years. By the mid 1970s, that trend had reversed itself. The four centuries between 1450 and 1850, the Little Ice Age, distort the long-term picture, as do a series of previous little ice ages – as defined by the spread of glaciers – some 3000, 5000, 8000 and 10 000 years ago.

FROZEN SANDS *Ice floes at Britain's coastal resort of Whitstable, Kent, in 1947 hint at extremes that may be either one-offs or part of a trend.*

Such predictions beg a question: what marks the onset of an ice age? It would take centuries of change for an ice cap to form on Ben Nevis or northern Canada, for the schools of North Sea herring to be driven south, for Russia's arctic ports to be permanently frozen. Since the idea of ice ages arose in Europe, one definition of an ice age is based on a European perspective. An interglacial is an interval of time during which oak and other deciduous trees are widespread in Europe – the demise of the oak signals the onset of an ice age, its return the end of one, as happened 10 000 years ago.

From the distribution of animals and plants, one trend emerges strongly. The year 5000 BC marked a high point in temperature. Then, certain species of oak and edible mussels flourished in Scandinavia. Since then, they have vanished. Elsewhere, vegetation belts have moved southwards and to lower elevations, and deserts have spread. As rock paintings in the Tassili N'Ajjer plateau show, the Sahara, now arid, was home to cattle farmers up until about 1200 BC. If this trend continues, global temperatures will reach ice-age levels – 6°C (10.8°F) cooler than now – in 18 000 years.

But Nature has a habit of casting spanners in her own works. The year 1816 was a disaster for Europe. It was the 'year without a summer', during which the price of oats doubled in England and the wine harvest in France was the latest on record. The reason was that in April the previous year, Tambora, a mountain on Sumbawa, Indonesia, had exploded in the greatest volcanic eruption of modern times. The explosion blew 3700 ft (1200 m) off the top of the mountain and blasted 35 cu miles (140 km³) of debris into the atmosphere. This shield of dust reflected enough sunlight to distort weather patterns worldwide for over a year. Dozens of volcanic eruptions since have done the same, if on a lesser scale.

One eruption would not change climates in the long term, but a series of eruptions would. One study analysing the climate over the last few millennia identified 18 phases of volcanic activity and relates them to the advance of glaciers.

THE HUMAN IMPACT

If Nature was left to her own devices, therefore, we should be in for another ice age in the not-too-distant future, on Earth's time scale. But she is not. Since the early

19th century, the Earth has been steadily pouring out debris of a different kind: industrial exhausts.

Over the past two decades, governments, magazines, newspapers, TV and radio stations, publishers, indeed people all over the world have either originated or been bombarded with information and theories about the impact of humanity on its environment. Dozens of words and phrases – environment, ecology, global warming, zero growth to name a few – acquired a currency previously unknown. For the first time, the world took official note that human success was bought at a price. Staggering advances

in medicine, agriculture and industrial growth all had their downside, producing numerous case histories of disaster. In Kazakhstan, millions of acres of newly ploughed lands turned to dust. In Japan, mercury effluent warped limbs. In the USA, pesticides killed lakes and destroyed the eggshells of birds. Everywhere, pressures grew to limit the dangers of development; and everywhere, equal and opposite pressures increased to intensify the same processes.

These issues were something entirely new in human history, and thus entirely new in the history of the Earth. For humanity they mattered intensely, for nothing could matter more than the future of the species. But for the living Earth, which has re-formed continents and atmospheres, and overseen the rise and fall of millions of species, what does the existence or non-existence of a featherless biped matter?

In the short term, the answer is: a good deal. Humanity has created another

element in the way the Earth regulates itself. We are now capable of re-engineering the planet, and already some of the implications are apparent.

Before the invention of agriculture, the human population was about 5 million. With each new phase in the development of technology – from farming to modern industry – the population increased ever more sharply. In the Earth's time scale, *Homo sapiens* has instantaneously multiplied 1000-fold. The world population has tripled since 1900, while energy consumption has risen tenfold. As the bestselling *The Limits to Growth* pointed out in 1972, such growth cannot go on because at some point, relatively soon, crisis will strike. In geological terms, the time scale of the coming crisis is minute. If present trends continue, a simple calculation shows that in 1600 years the mass of humanity will equal the mass of the Earth. Clearly, long before that point, on a time scale that may be decades or may be centuries, human beings will run out of room, food, fuel, raw materials and/or the ability to dissipate its waste products.

The latter, in particular, threatens the Earth's ancient chemical stability by releasing huge amounts of gases previously stored in the ground and in plants. Take carbon dioxide, a 'greenhouse gas' that helps to keep the Earth warm and has been maintained at low levels by the action of plants. Without the plants, or with the addition of more carbon dioxide into the atmosphere, the gas would rise again. Indeed, it now is. Forest destruction releases some 2 billion tons of carbon dioxide into the air each year (an area of forest the size of Great Britain is cut down every year, a rate which, if unchanged, will lead to the disappearance of all tropical forests by the middle of the next century). Simultaneously, industrial emissions of carbon dioxide rise; they increased from 2 billion tons annually to over 5 billion between 1960 and 1990. During the

Atmosphere

NEGATIVE FEEDBACK:
As the world heats up, more water evaporates causing more clouds to form. These reflect some of the Sun's heat, reducing the effect of greenhouse gases.

Clouds

TIPPING THE BALANCE
Some of the results of the increase in greenhouse gases reduce the effects of those gases – a process known as negative feedback. Other results increase their effects – known as positive feedback. Scientists do not yet know whether a balance will be established between the two.

Carbon dioxide produced by burning coal

Forests are natural 'sinks', soaking up greenhouse gases. Deforestation reduces the size of these natural sinks

Coal-fired power station

Carbon dioxide from burning petrol

Carbon dioxide produced by burning wood

Plants soak up some carbon dioxide

NEGATIVE FEEDBACK:
More carbon dioxide means that more plants thrive and therefore more carbon dioxide is soaked up

POSITIVE FEEDBACK:
Water vapour is a greenhouse gas. Warmer temperatures put more water vapour into the air

Farms produce methane (from animals) and nitrous oxide (from fertilisers)

Plankton in the sea absorb some carbon dioxide

NEGATIVE FEEDBACK:
More carbon dioxide helps more plantkton to grow, soaking up more carbon dioxide

Melting ice caps results in a reduction in the area that reflects heat from Sun

EXPANDING CITIES *Urban
growth, as in this shanty town on
the edge of Rio de Janeiro, is both
a cause and a consequence of the
boom in the human population.*

20th century, the amount of carbon dioxide
in the atmosphere has risen from 280 parts
per million (0.028 per cent), to about 305
in 1960, and 350 in 1990. The current growth
rate is 0.5 per cent a year, and the atmospheric content of carbon dioxide is expected to double, even if controls are imposed.

Carbon dioxide is not the only greenhouse gas that is increasing in volume in the
Earth's atmosphere. Methane is also on the
increase. The quantity in the atmosphere
doubled between 1960 and 1990 and is increasing at 1 per cent a year. Nitrous oxide,
mainly from nitrogen fertilisers, is also rising.

The effect of this may, on balance, far
outweigh the effects of any cooling trends.
As a rule of thumb (though there is no consensus), many scientists accept that a dou-
bling of carbon dioxide would result in a
1°C (1.8°F) rise in average temperatures,
with a more pronounced effect in higher
latitudes. This is not a simple cause-and-
effect, because greater heat also initiates a
contrary cycle: greater evaporation, more
clouds, more sunlight reflected, more cooling. But overall, by the end of the 21st century, temperatures worldwide could be 3°C
(5.4°F) higher, with extensive changes
being imposed by shifting patterns of agriculture and sea levels several inches higher
than at present.

In the medium term, in the words of
one expert, J. Murray Mitchell, it is likely
that consumption of the bulk of the world's
known fossil-fuel reserves would plunge the
planet into a 'super-interglacial age', a millennium during which most of the Greenland and Antarctic ice caps would melt,
'raising sea levels around the world enough
to submerge many of our coastal population centres and much productive farm-
land'. This consequence, though, might be
offset by renewed fertility in the arid zones
of North Africa and the Middle East. By
then, all the stored fossil fuel would have
been expelled into the air.

It would take the Earth another few centuries to re-absorb the gas. After 2000 years,
the ancient underlying cycles would re-
assert themselves, and the Earth would resume its delayed plunge into the next ice
age, as dictated by the iron laws identified
by Milankovitch. In 23 000 years from now,
the Earth would once again find itself in
the depths of a new ice age.

All this involves regular but rapid
change. However, a slow and steady increase in temperature could also involve a
change that is anything but slow and steady.
The key player in this scenario is the West
Antarctic ice sheet, a peninsula that makes
up 11 per cent of all the ice on Antarctica.
There is some evidence to suggest that the
whole sheet is unstable. If rising temperatures make it move more easily, it could
slide en masse into the ocean. The sudden
addition of 350 000 cu miles (1.4 million
km^3) of ice would have rapid and profound
effects: warming of the Antarctic currents,
loss of the ice sheet's mirror-like ability to reflect sunlight, more warming, and within a
year or two a rise in sea levels of 23 ft (7 m).

If the collapse kick-started a total melt-down, the temperature rise would continue for centuries, a speeded-up version of the 'super-interglacial' into which we may already be heading.

LIFE AS A PRODUCT OF THE UNIVERSE

Life on Earth, then, has had a drastic effect on its home, mainly by remodelling the atmosphere, first to make it fit for life by exchanging the carbon dioxide for oxygen; then, after the evolution of intelligence, pumping the carbon dioxide back out again into the air from which it was absorbed in the first place. Eventually, our species may either so foul its own nest as to destroy itself, or find some way to continue its expansion. Possibly, the only way to do that will be to leave home. Given that the very nature of life involves adaptation in order to continue, and given the outward urge that seems to be part of the human character, that seems the more likely. Indeed, the technology that would enable mankind to escape from Earth already exists. The Earth, the womb of life, could be on the point of seeding the Universe from which it sprang.

When the space programmes of several nations developed, from the 1960s onwards, it became common-place to see Earth-based rock-etry as opening a way to the future of space travel. In fact, experience has shown that this is not the way forward. It is too expensive to manage a space programme from the surface of the Earth; and instruments can do a better job of gathering information. To make space exploration and colonisation work, it must offer soundly based economies and new freedoms to those who embark upon the venture. This can

only be done by establishing self-sustaining colonies, an idea that was explored by the US physicist Gerard K. O'Neill in his book, *The High Frontier*.

O'Neill based his proposals on two propositions: that space offers an almost infinite source of power – the Sun – and that there is enough in the way of raw materials on the Moon or asteroids to build afresh in Earth's orbit. He explored the idea of space colonies, whole cities in which 10 000 people or more would live on the inside surfaces of sealed spheres or cylinders, which would spin to provide gravity. Powered by the heat of the Sun, mining their raw materials cheaply from low-gravity sources such as the Moon, these colonies would grow their own food and be self-sustaining. Much of the metallic material for building is available on the Moon, once a mining operation was established there. Asteroids could provide other elements, such as carbon, nitrogen and hydrogen.

Once in place, the first colonies would, in the course of generations, build newer, larger ones, housing 100 000 people, or

even 1 million people. O'Neill imagined that as conditions on Earth became harsher, there would be a steady drift of colonists into space, financing their one-off emigration much like the emigrants to the United States from Europe in the 19th century. At first, the colonies would be part of the Earth-Moon system, but there is no reason why they should not be almost anywhere in the Solar System. Since they would not need the explosive acceleration needed to escape the Earth's gravity, the colonies could be accelerated gradually, shifting orbit over many years.

These may seem like science-fiction scenarios, but considerable work has been done on what would be needed to develop and sustain such colonies, with the conclusion that current technology could serve perfectly well. The only thing that is missing is the will to initiate a revolution that would mark a new stage in the evolution of humanity. The first colony might take decades to build, but once it was built productivity would accelerate. Within a century, there could be dozens, and within a

RIVER OF DEATH
A Tasmanian river dies,
polluted by copper mining
near Queenstown.

1000 years, mankind would be a solar, rather than purely a terrestrial, animal.

O'Neill's theories, whether or not they are ever realised, lead on to a topic that forces another revision of the Earth's place in the Universe. It seems that the Earth has not only produced life, but has also acquired the capacity to seed its space environment with life. Is this something unique in the history of the Universe?

LIFE ELSEWHERE

To put the question in another way: if the Earth can do this, and if the Universe is uniform, and if Sun-like stars can produce Earth-like planets, and if such planets can produce intelligent life, and if intelligent life can escape from its home planet and its home solar system in the course of only a few thousand years, where is everybody?

There are a lot of 'ifs' here. But astronomers are sure that the Universe is uniform, so sure that a widespread assumption has arisen that there is – there must be – life elsewhere. This assumption has deep roots, comparable to the contrary assumption that God made Man in his own image as the unique centre point of Creation. As religion retreated and science advanced, as mankind became relegated to ape-hood and the Earth to a galactic suburb, the idea of extraterrestrial life became a commonplace first of philosophy, and then of science. The American astronomer Percy Lowell's observations – he thought he saw 'canals' on Mars and claimed that they were built by intelligent life forms – were the result of wishful thinking, but the discovery that amino acids could be produced from the Earth's earliest constituents and that complex molecules existed between the stars suggested that life was indeed universal. As a result, a great deal of money has been expended by many nations on the Search for Extraterrestrial Intelligence, SETI for short.

The first attempt to give theory a semblance of reality came in the 1970s, when the US astronomer Frank Drake provided a rough-and-ready statistical approach to the problem. He equated seven factors basic to the existence of intelligent, communicative life in the Universe: the rate of star formation, the fraction of stars with planets, the

A MESSAGE FROM EARTH

In the 1970s two probes, Pioneer 10 and 11, flew past Jupiter, and then headed out of the Solar System into interstellar space. Though it will take them several million years to cross the void even to the nearest stars, the probes offer an opportunity to send a physical message from Earth. An engraved plaque was attached to each probe on the off-chance that some alien civilisation will find them and want to know about their origins.

Devised by the American astronomer Carl Sagan, the message is written in what should in theory be a universal language: science. Any civilisation capable of finding it should be able to understand it.

It shows, on its top left corner, a diagram of a hydrogen atom, which produces radiation common throughout the Universe. The frequency of the radiation provides a suitable unit to measure both space 8 in (20 cm) and time 1420 MHz. Below that, a starburst pattern maps 14 objects, fast-spinning pulsars,

as seen from the Sun, with their distance suggested by the different lengths of line and their spin-times indicated in binary code, using the hydrogen atom as the basic unit.

Any civilisation clever enough to intercept the probe would know about pulsars, and could thus identify both them and the diagram's central star group. And since pulsars slow with time, the aliens could also work out when the plaque had been dispatched, by comparing the difference between the pulsars' specified frequencies and the ones current millions of years hence.

At the bottom is the Solar System, a unique combination of planets that would be enough to define our Sun among its neighbours. The diagram of the Pioneer specifies the planet of origin. A scale on the far right gives the height of a woman in binary – 8 x 21 cm (168 cm/ 5 ft 6 in) – who, along with her companion, is drawn in scale to the Pioneer probe itself.

The two creatures may be the most puzzling element in the whole design, and the male's gesture of peace the most puzzling thing of all. But their very oddity should suggest both biological and cultural interpretations. If the plaque inspires a research trip, though, the world found by the curious alien astronauts will be very different from the one recorded here.

LETTERS TO ALIEN CULTURES
The duplicate plaques (below left) on the Pioneer probes (below) could still be drifting between the stars long after human life on Earth has died.

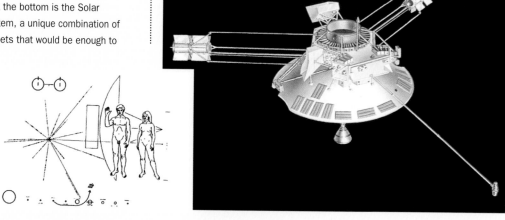

SPACE VOYAGER
The international Freedom space station could become a steppingstone to other planets, or even other stars.

number of planets suitable for life, the percentage of planets that actually produce life, the likelihood of intelligent life, the chances of the desire to communicate, and the longevity of the civilisation. One estimate that emerged – mostly guesswork, and of more historical than practical significance – was that there should be a million or more advanced civilisations in our Galaxy alone. It sounds a lot, but they would be well scattered, with one civilisation for every 100 000 stars. We would probably have no near neighbours.

In the next two decades astronomers acquired a few more facts. They know a good deal about star formation. There are about 10 billion Sun-like stars in our Galaxy. But

how many have planets? Stars are all too far away for scientists to detect planets directly but they can be inferred from a star's motion. In 1956, after 20 years of study, Peter van de Kamp of Sproul University, Pennsylvania, revealed that a faint red dwarf, Barnard's Star, had a wobble that was possibly induced by the presence of two large planets. And in 1995 and 1996, three other stars were found to wobble. In the intervening 40 years, techniques had improved and the conclusions were more definite.

These stars – numbered according to their constellation 51 Pegasi, 47 Ursae Majoris and 70 Virginis – had a planet each, very large ones, ranging from half the size of Jupiter to six times its size. They could hardly be inhabited, but they were planets nevertheless – or planet-*like*, in any event. One planet (the one belonging to 51 Pegasi), though massive, is in an orbit as tight as Mercury's; another (the one round 70 Virginis) is in an egg-shaped orbit. Neither

fits easily into theories of planetary origins. However, where there are four such worlds, there are sure to be more. Around the turn of the millennium, the search will be on for Earth-like planets, using sensing devices in Earth-orbit.

If intelligent life exists, why is it not manifest? Although the idea of an interstellar conversation may be ludicrous – it would take eight years simply to exchange a 'Hello' with a planet orbiting the *nearest* star – it surprises some people that in 35 years of searching, astronomers have not identified a single artificial signal. If life is common, then some intelligent, broadcasting, space-travelling civilisation should surely have preceded us by enough time to have left electronic traces. Earth pumps out signals strong enough to be detectable across the Galaxy, so why has no one else done the same?

The question, and the notable absence of an answer, suggests a disturbing and

depressing line of thought. The idea of alien intelligences has now become a strand in Western culture, an inspiration for millions, a replacement, perhaps, for a vanishing religious faith, a reassurance that 'we are not alone'. Yet what of the great silence? What if we are alone? Or what if we are the first intelligent life? What responsibilities does that confer? And how does it recast our view of our home, the Earth?

NEMESIS

There is another, equally disturbing, thought suggested by the silence of infinity. Perhaps there have been other great civilisations out there. Perhaps they have simply not survived, not because they destroyed themselves in nuclear wars or sinks of pollution, but because the Universe destroyed them, by

the same means that it destroyed the dinosaurs. Over an extended time scale the Universe is a dangerous place.

Though a major impact hardly ranks very high as a worry for most people, astronomers have listed hundreds of asteroids and comets crossing the Earth's orbit. They expect to find thousands more. In time, one will hit. No known rock is on a collision course with Earth, and all that astronomers can go on is samplings of the past. An impact as big as the one that struck Tunguska, Siberia, in 1908, might happen once a century, while bodies up to a mile or so across might strike once every few hundred thousand years. Possibly, if the Permian and Cretaceous extinctions were both caused by impacts, little wandering planets up to 6 miles (10 km) across might strike every 150 million years or so. But even a 'little' strike by an asteroid a mere 350 yd (320 m) across would release more energy than is stored by all the world's nuclear weapons.

These risks can be quantified statistically. A report by the

SEARCHING THE SKIES *Whole galaxies, such as NGC 253, can be monitored for artificial signals by radio telescopes.*

US Congress in 1992 estimated that an impact with a 1 mile (1.6 km) wide asteroid, causing worldwide devastation, might cause the death of one quarter of the world's population. Such an asteroid might put nearly 1000 times more dust into the air than Mount Pinatubo did in 1991, an eruption that cooled the Earth measurably for a year. If one of these objects strikes once in 500 000 years, there is an annual risk to each of us of 1 in 2 million. In actuarial terms, everyone on Earth has a 1 in 30 000 chance of dying by asteroid strike.

Another problem is how to clothe statistics with hard fact. Astronomers have a particular interest in the Taurid meteor shower that arrives every June 28 (it is not a spectacular one because the shower occurs in daylight). The specks of dust lie in the same orbit as a comet, Encke's comet, which circles the Sun relatively closely within the confines of our Solar System. But there is more than dust in Encke's orbit. It is also associated with ten asteroids, and at least one astronomer, Duncan Steel of the Anglo-Australian Observatory, theorises that the dust, the asteroids and the comet are all derived from a single large object that fell into the Solar System some 20 000 years ago. He places the risks of major impacts higher – a Siberia-like strike every 50 years, a 1 mile (1.6 km) wide one every 100 000 years. One scientist, Victor Clube of Oxford, equates the end of ancient civilisations with a rain of destruction from the skies.

The ammunition is certainly out there. Way beyond the range of visibility lies the Oort Cloud, with its store of comets by the billion. Closer in, the larger bullets are visible: some 5500 asteroids have been listed so far, with more being discovered every year. Several large ones are in eccentric orbits that swing close to the Earth – Icarus careened past in 1968 at a distance of 4 million miles (6.4 million km) – and over 1000 'near-Earth' asteroids over $2/3$ mile (1 km) across have been recorded. The outer Solar System contains many bodies up to 125 miles (200 km) across, in orbits that could be destabilised by the passing of Jupiter, with consequences seen when Shoemaker-Levy broke up and struck Jupiter itself in

July 1994. Also in 1994, James Scotti, part of the US Spacewatch programme set up at the University of Arizona's Steward Observatory, recorded an asteroid passing only 65 000 miles (104 600 km) away.

Several observatories have asteroid-watching programmes, and should in years

GLOBAL IMPACT

When the volcanic island of Krakatau in Indonesia exploded in 1883, 1½ cu miles (6 km³) of rock and ash were blasted 30 miles (50 km) into the air. The region was in darkness for 2½ days, and dust drifted several times around the Earth, creating spectacular sunsets throughout the following year. Tidal surges were recorded as far away as South America and the English Channel.

to come be able to identify and coordinate their information. In the words of one researcher, Tom Gehrels, Professor of Planetary Science at the University of Arizona, Tucson: 'If there is an asteroid out there with our name on it, we should know by the year 2008.'

INCOMING MISSILE *The comet Swift-Tuttle, which swung past the Sun in 1992, is due to return in 2126 in an orbit so close to the Earth that it could hit.*

All of this has great relevance to the history of the Earth. With, say, ten years or more warning, humanity, and the Earth itself, could be in a new era of evolution. For the first time the Earth has a means of self-defence, for scientists have already proposed the funding required not simply to keep watch, but to take action if necessary. As with the building of space colonies, the technology – in the form of rockets and explosives – already exists to blast or guide an asteroid from its destructive path. According to Gehrels, the chances are that we would have a century to ward off any big strike, and 'given that time, a modest chemical explosion near an asteroid might be enough to deflect it. The explosion will need to change the asteroid's trajectory by only a

small amount so that by the time the asteroid reaches the Earth's vicinity, it will have deviated from its original course enough to bypass the planet.'

As happened before in its history, when early organisms helped to create a life-supporting atmosphere, the Earth has produced a life form that can help to ensure the Earth's own future. This time, if it happens, it will be done by the exercise not of biochemical forces but of the most influential quality to emerge from those forces: consciousness.

THE DRIFTING CONTINENTS

All of the above – climatic change, ice ages, human impact, asteroid impact (or avoidance) – will take place on the briefest of time scales, in terms of Earth's whole life. Throughout those changes, other imperceptible changes will be occurring, ones that will only become measurable on a very much longer time scale. The continents remain in motion, the sea floor is always spreading, volcanoes erupting, the Earth quaking. There is no reason to think that current trends in plate tectonics will suddenly become unstable, and

IMPACT ZONE *Meteor Crater, Arizona, is evidence that Earth is subject to occasional large-scale bombardment.*

scientists can predict with a fair degree of certainty what the face of the Earth will look like in 50 million years from now.

North and South America are swinging anti-clockwise, closing the South Atlantic and opening the North. North America will continue grinding past the Pacific plate along the San Andreas Fault. The Great Rift

Valley, the depression running south from the Jordan valley down through Kenya to Mozambique, will tear open, splitting East Africa, Arabia and places north of them away from the rest of Africa and Europe, and opening a new ocean into the Mediterranean. Siberia will again be free, forming the northern wing of the landmass that in-

cludes Arabia and India. Australia will continue its drift northwards. Only Antarctica will be stable, locked over the South Pole, a focus for growing and shrinking ice caps.

In terms of human history, Gaia herself will seek to retain the stable personality she has developed in the course of her life. Whether she will be allowed to do so is an open question. There is another image putting her age into perspective: if you imagine Earth's history as an outstretched arm, then human history is a fingernail. And fingernails can do sudden and catastrophic damage to a beautiful face.

THE END OF THE END

There is a story by the science-fiction writer Arthur C. Clarke about a spaceship exploring a distant part of the Galaxy. It approaches the dead stump of an ancient star and finds, circling at a remote distance, a tiny, burned-out ember of rock. The astronauts find it because it is emitting a radio signal powered by a radioactive source that will last for billions of years. The signal guides them to a site deep inside the rock. There they find a cache of information that tells what happened. The star once had a planet brimming with intelligent life. The star had used up its fuel, contracted, exploded into a supernova and consumed the planet. Before that happened, the civilisation that occupied the planet, knowing that its death was not the end of all life, placed the cache of information to tell the future of a vanished past. The visiting astronauts perform calculations. They discover when the star died, and how brightly it would have shone in Earth's skies. Its light would have reached the Earth some time around the time of Jesus's birth. The star that killed off this rich and admirable planet had been the Star of Bethlehem.

In this fictional parable about the cycle of life, death and rebirth, Clarke makes an unstated parallel. If new life, in the form of

SPLITTING CALIFORNIA *The San Andreas fault acts as a reminder that California is unstable and will in the future look very different.*

inspiration, can spring from the death of a distant star, why should the Sun and Earth not find continued if shadowy existence in some comparable way?

Consideration of what that might mean involves a journey from the present into the distant future, when the Earth faces death, as all the planets and the Sun itself will. For the Earth, there are many possible conclusions. One possibility, of course, is that of a sudden, fiery end brought about by impact with some wandering object so enormous that the Earth would be shattered. Set this aside: no such impact has yet taken place in the Solar System in recent aeons because in the process of creation all likely candidates were blotted up to form the planets themselves. The most likely end, the one that astronomers understand as an inevitable consequence of processes that have been in existence since Earth's beginning, ties the fate of the Earth to that of the Sun.

Some 5 billion years hence, the Sun – a normal star whose destiny is foreshadowed by the fate of countless similar stars – will have burned up its hydrogen core, creating a core of helium. Slowly, this core will grow, and then contract, adding to the heat generated by the remaining hydrogen burning in a shell around it. In response, the Sun will expand. It will become a 'red giant', blowing up like a balloon in a tremendous surge of energy until it spreads out beyond the orbits of Mercury and Venus.

This will be the end of any life on Earth, for the Sun will now be hundreds of times its present size, with twice the surface temperature. The Earth will be orbiting inside the Sun's outer atmosphere and the interaction will be catastrophic, ripping off the Earth's atmosphere and vaporising its mantle. To survive, the Earth would have to be transported to a safe distance. But, even supposing our descendants should have the will and the technology to do that, they would merely be exchanging one grandstand view of disaster for another.

The hydrogen in the Sun's outer layers will be used up, and the helium core will shrink until it, too, undergoes nuclear fusion, burning helium to carbon in a reaction that will release ten times as much

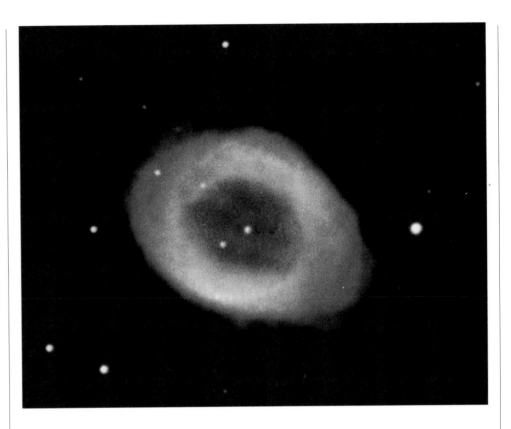

energy as hydrogen fusion did. The sequence will then be repeated – the carbon core growing while the helium shell burns away, expanding into a new red-giant phase. This is a time of increased instability, with the Sun bubbling in pulses lasting a few years or tens of years, and eventually it will cast off its outer mantle in a sort of smoke ring, known as a planetary nebula (because William Herschel thought the first one he saw was a planet). A few hundred thousand years later, the outer shell will dissipate in the interstellar void, leaving behind a carbon core that has insufficient mass to contract and initiate yet another type of nuclear reaction. It will become a glowing 'white dwarf', a small one the size of the Earth, but extremely dense – a spoonful of white dwarf would weigh as much as a car. White dwarfs of this size have no active future ahead of them (it is the bigger ones that collapse still further, eventually forming black holes). For billions of years, the Sun, or what remains of it, will simply radiate its heat away towards invisibility as a 'black dwarf', a dying ember that will do nothing more than give up its heat,

OUR SUN'S FUTURE *The Ring Nebula is the gaseous remnant of a Sun-like star that blew away its outer mantle some 20 000 years ago.*

for ever, gradually solidifying into a crystal of incredible rigidity, invisible against the blackness of space.

Perhaps, even then, a cinder of rock will remain in orbit around the crystal, burned beyond recognition by the fires of the red giant, its complex of minerals reduced to an elemental flux and carried off into interstellar space. In nearby corners of the Galaxy, meanwhile, other stars will have formed, drawing in the scattered tendrils of the Sun's envelope and using them to build themselves and their own sets of planets. Perhaps on one such planet an equally rich variety of materials as on Earth will form, in equal stability, eventually producing intelligent life that will explore its neighbourhood and find the dense crystal of rock with its circling ember. And visiting astronauts will know that the little circling cinder was once a crucible for life and a home.

PICTURE CREDITS

3 SPL/STSI. 6 SPL/NASA. 7 SPL/Tom Van Sant/GeoSphere Project. 8 The Bridgeman Art Library/British Library, TL; SPL/NASA, BR. 9 SPL/US Geological Society, TL; SPL/Gordon Garradd, BR. 10-11 Siena Artworks Ltd, London/Lee Peters. 12 SPL/NASA. 13 Mary Evans Picture Library, TL; SPL/Royal Observatory, Edinburgh, BR. 14 SPL/Tony Hallas. 15 SPL, BC; Siena Artworks Ltd, London/Lee Peters. 16 Siena Artworks Ltd, London/Lee Peters. 17 SPL/Mullard Radio Astronomy Laboratory, TR; Siena Artworks Ltd, London/Lee Peters. 18 SPL/A. Barrington Brown, TL; Mary Evans Picture Library, BC; Sienna Artworks Ltd. London/Lee Peters. 19 SPL/NASA, TR; Siena Artworks Ltd, London/Lee Peters. 20 Science & Society Picture Library, BL; TC; Mary Evans Picture Library, CR. 21 Science & Society Picture Library, TL: SPL/David Parker, BC; TR; SPL/Peter Menzel, BL; SPL/Dr Jean Lorre, CL; SPL/Royal Observatory, Edinburgh, CR; SPL/Roger Ressmeyer, BR. 22 Camera Press/Charles Green, TL; Corbis-Bettmann, BR. 23 Siena Artworks Ltd, London/Lee Peters. 24 SPL/Mehau Kulyk. 25 SPL/Royal Observatory, Edinburgh. 26 SPL/NASA. 27 SPL/STSI. 28 SPL. 29 SPL/NASA. 30 SPL/NOAO. 31 SPL/Royal Observatory, Edinburgh, TR; Siena Artworks Ltd, London/Lee Peters. 32 Siena Artworks Ltd, London/Lee Peters. 33 SPL/Jodrell Bank. 34 SPL/NRAO, TL. 34-35 SPL/STSI. 36 SPL/STSI. 37 SPL/Max Planck Institute fur Extraterrestrische Physik, BR; Siena Artworks Ltd, London/Lee Peters. 38 SPL/Royal Observatory, Edinburgh. 39 SPL/STSI. 40 SPL/STSI, BL; Siena Artworks Ltd, London/Lee Peters. 41 SPL/Nick Sinclair, TL; Siena Artworks Ltd, London/Lee Peters. 42 SPL/Royal Observatory, Edinburgh. 43 SPL, TL; BR. 44 SPL/Gordon Garradd. 45 The Bridgeman Art Library/British Library, BR; Siena Artworks Ltd, London/Lee Peters. 46 SPL/NASA, TL; Siena Artworks Ltd, London/Lee Peters. 47 SPL/NASA. 48 SPL/MSSSO/ANU, TL; SPL/Harrington et al, BL;SPL/STSI. 49 SPL/JISAS/Lockheed. 50 Siena Artworks Ltd, London/Lee Peters. 51 Siena Artworks Ltd, London/Lee Peters. 52 Michael Holford, TR; Ann Ronan Picture Library/Image Select, BL. 53 SPL/NASA, TR; BL. 54 The Natural History Museum, London, BC; Siena Artworks Ltd, London/Lee Peters. 55 SPL/Ian Steele & Ian Hutcheon, BL; SPL/NASA, TR. 56 SPL/NASA. 57 SPL/NASA. 58-59 SPL/David P. Anderson/SMU/NASA. 59 Siena Artworks Ltd, London/Lee Peters, TR. 60 SPL/US Geological Survey. 61 SPL/NASA, TR; Siena Artworks Ltd, London/Lee Peters. 62-63 SPL/NASA. 64 SPL/NASA, TL;TR; Siena Artworks Ltd, London/Lee Peters. 65 SPL/NASA, TC; BR. 66 SPL/NASA, TL; BR. 67 SPL/John Sanford, TL; SPL/NASA, BR. 68 The Bridgeman Art Library/Church of St Madeleine, Troyes, BL; Siena Artworks Ltd, London/Lee Peters. 69 GeoScience Features. 70 Mary Evans Picture Library, TR; Trinity College, University of Dublin, BC. 71 Ann Ronan Picture Library/Image Select, TL; SPL/David Parker, TR; Siena Artworks Ltd, London/Ed Stuart. 72 Tony Waltham, BL; BR; Siena Artworks Ltd, London/Ed Stuart. 73 Planet Earth Pictures/Jim Brandenburg. 74 SPL/Los Alamos National Laboratory/P. Roberts/UCLA, TR; Siena Artworks Ltd, London/Ed Stuart. 75 Siena Artworks Ltd, London/Ed Stuart. 76 Siena Artworks Ltd, London/Ed Stuart. 77 SPL/Earth Satellite Corp. 78 SPL/NASA. 79 Scala/Biblioteca Nazionale, Firenze. 80 SPL/NASA, TR; BL. 81 SPL/NASA. 82 Siena Artworks Ltd, London/Ed Stuart. 83 SPL/NASA. 84 Siena Artworks Ltd, London/Ed Stuart. 85 SPL/David Parker, TL; Siena Artworks Ltd, London/Ed Stuart. 86 SPL/S. Clement/Publiphoto. 87 Siena Artworks Ltd, London/Lee Peters. 88 Siena Artworks Ltd, London/Lee Peters. 89 AKG Photo, London, TL; Natural History Museum, London, BR. 90 SPL/Simon Fraser. 91 Planet Earth Pictures/Adam Jones, BL; Wilderness Photographic Library/John Noble, TR. 92 Siena Artworks Ltd, London/Mick Saunders. 93 BCL/Jules Cowen, TL; SPL/Dr Jeremy Burgess, BL. 94 Mats Wibe Lund, B; Mats Wibe Lund/Sturla Fridriksson, TR; Mats Wibe Lund/Gardar Palsson, BL. 95 BCL/Gerald Cubitt, TL; SPL/David Scharf, BR. 96 Mary Evans Picture Library, TR; Planet Earth Pictures/Jonathan Scott CR. 97 SPL/NASA, TL; GeoScience Features, BR. 98 Wilderness Photographic Library/John Noble. 99 SPL/NASA. 100 The Bridgeman Art Library, TR; OSF/Ronald Toms, BR. 101 DRK Photo, TR; OSF/Ronald Toms, BL. 102 SPL/Sabine Weiss. 103 BCL/Atlantide, TL; DRK Photo, BR. 104 Images Colour Library. 105 SPL/Johnny Autrey. 106 NHPA/A.N.T. 107 Tony Stone Images. 108 SPL/Dr Kari Lounatmaa. 109 Planet Earth Pictures/I & V Krafft/Hoa Qui. 110 SPL/Sinclair Stammers. 111 OSF/Daniel J. Fox, TC; DRK Photo/Kim Heacox, B. 112 DRK Photo/Tom Bean. 113 DRK Photo/Larry Ulrich. 114 SPL/BP/NRSC. 115 Ardea/Françoise Gohier. 116 The Natural History Museum, London, BC; Siena Artworks Ltd, London/Andrew Thompson. 117 Siena Artworks Ltd, London/Andrew Thompson. 118 Tony Waltham. 119 Tony Stone & Associates/Jim Sparks. 120 SPL/University of Cambridge. 121 SPL/Dr Gopal Murti, TL; Simon Conway Morris/University of Cambridge. 122 SPL/Sinclair Stammers, BL; DRK Photo/John Cancalosi, BR. 123 Siena Artworks Ltd, London/Mick Saunders. 124 SPL/Profs. P.M. Motta, S. Makabe & T. Naguro. 125 OSF/Stan Osolinski, BL; OSF/Kattie Atkinson, TR. 126 Ardea/D. Parer & E. Parer-Cook, BL; NHPA/Bill Wood, TR. 127 Hedgehog House/Colin Monteath. 128 GeoScience Features, TL; The Natural History Museum, London, BR. 129 Ardea/Ferrero-Labat. 130 Chris Pellant, B; Images Colour Library, CR. 131 SPL/Simon Fraser. 132 Simon Conway Morris/University of Cambridge. 133 GeoScience Features, L; OSF, CR. 134 Simon Conway Morris/University of Cambridge, TC; BL; BR. 135 GeoScience Features, TL; The Natural History Museum, London,BR. 136 Ardea/François Gohier, TR; Siena Artworks Ltd, London/Jim Robins. 137 OSF/Michael Fogden. 138 SPL/Sinclair Stammers, TL; BCL/Jens Rydell, BR. 139 SPL/Sinclair Stammers. 140-1 Ardea/Françoise Gohier. 142 OSF/Breck P. Kent, TR; SPL/Peter Menzel, B. 143 BCL/John Cancalosi. 144 Popperfoto. 145 SPL, TL; SPL/Laboratory for Atmospheres/NASA/GSFC, TR. 146 Robert Harding Picture Library, TL; Ardea/J. Swedberg, BL. 147 Siena Artworks Ltd, London/Mick Saunders. 148 NHPA/Daniel Heuclin. 149 Ardea/Jean-Paul Ferrero. 150 Science Photo Library/NASA, BC; SPL/David Hardy, BR. 151 SPL/David Hardy. 152 Corbis-Bettmann, TC; SPL/Luke Dodd, BL. 153 SPL/Rev. Ronald Royer, TR; Images Colour Library, BL. 154 SPL/David Parker. 155 SPL/NOAO.

FRONT COVER: SPL/Royal Observatory, Edinburgh; SPL/Earth Satellite Corp., C